Machine Learning Integration in Power Electronics

Hasan Ali Gamal Al-kaf · Kyo-Beum Lee

Machine Learning Integration in Power Electronics

Essential Principles, Implementation Practices, and Industry Applications

Springer

Hasan Ali Gamal Al-kaf
Department of Electrical and Computer Engineering
Ajou University
Suwon, Korea (Republic of)

Kyo-Beum Lee
Department of Mobility Engineering, Department of Electrical and Computer Engineering
Ajou University
Suwon, Korea (Republic of)

ISBN 978-981-95-3844-7 ISBN 978-981-95-3845-4 (eBook)
https://doi.org/10.1007/978-981-95-3845-4

This Springer imprint is published by the registered company Springer Nature Singapore Pte Ltd.
The registered company address is: 152 Beach Road, #21-01/04 Gateway East, Singapore 189721, Singapore

Preface

The integration of machine learning into power electronics is rapidly transforming the field, enabling advanced control, optimization, and predictive maintenance strategies that were previously unattainable. As power electronic systems become more complex and demanding, traditional control and diagnostic methods face significant challenges in terms of computational efficiency, adaptability, and real-time implementation. Machine learning offers promising solutions by leveraging data-driven approaches to enhance control precision, reduce energy losses, and improve system reliability.

This book delves into the integration of machine learning with power electronics, with a focus on real-time applications, motor drives, inverters, and fault detection systems. It presents a comprehensive study of various machine learning techniques, their implementation challenges, and the opportunities they provide for power electronics engineers.

The early chapters introduce power electronics fundamentals and highlight the role of machine learning in design optimization, control, and predictive maintenance. Subsequent chapters delve into specific machine learning algorithms, including supervised, unsupervised, and reinforcement learning, as well as their implementation in various power electronics applications. This book also discusses practical implementation strategies, including real-time deployment on digital signal processors (DSPs). In addition, case studies and experimental validations are included to demonstrate the efficacy of machine learning methods in improving the performance of permanent magnet synchronous motor (PMSM) drives, multilevel inverters, and fault detection mechanisms. Later chapters discuss recent studies that implement machine learning in real time, providing insights into the best practices and recommendations. A strong emphasis is placed on the real-time implementation of advanced machine learning techniques, including deployment on FPGAs and other microprocessors. This book also summarizes experimental setups that facilitate machine learning applications, and examines recent algorithms such as physical informed neural networks, explainable machine learning, and low-complexity neural network models. Additionally, it explores solutions to key challenges, including the handling of unbalanced and limited datasets, ensuring computational efficiency, improving system adaptability, and addressing the uncertainty of machine learning methods in power electronics.

This book is intended for researchers, engineers, and students seeking a deeper understanding of how machine learning can be leveraged to enhance power electronics. By providing both theoretical insights and practical guidelines, it aims to bridge the gap between machine learning innovations and their real-time applications in power electronic systems.

This work is intended to serve as a valuable resource for those looking to harness the potential of machine learning in advancing the field of power electronics.

Suwon, Korea (Republic of)

Hasan Ali Gamal Al-kaf
Kyo-Beum Lee

Contents

Chapter 1
Introduction to Machine Learning (ML) in Power Electronics

Abstract This chapter introduces the field of power electronics and explores the growing role of machine learning (ML) methods in this area. Power electronics focuses on the efficient conversion and control of electrical energy through semiconductor devices, playing a critical role in a wide range of applications, including industrial automation, renewable energy systems, and electric drives. As power converter systems become increasingly complex and dynamic, conventional control and design methods often struggle to manage their nonlinear and time-varying behaviors. In response to these challenges, ML has gained significant traction for its ability to model and adapt to complex system dynamics using data-driven approaches. Recent advancements in ML algorithms and computational power have enabled more effective handling of the intrinsic complexity of power electronics. This chapter highlights three core areas where ML integration has shown considerable impact: design optimization involves selecting topologies, sizing components, and synthesizing circuits more efficiently. Intelligent control enables adaptive and robust real-time operation while predictive maintenance ensures system reliability by supporting condition monitoring, fault diagnosis, and remaining useful life prediction.

1.1 Overview of Power Electronics

Power electronics is a specialized field within electrical engineering that focuses on the efficient conversion, control, and management of electrical energy. It facilitates the transformation of power to meet the specific requirements of various applications, ranging from consumer electronics to large-scale industrial systems. This field is instrumental in improving energy efficiency and enabling the integration of renewable energy sources into modern grids. At the heart of power electronics are semiconductor devices such as diodes, thyristors, MOSFETs, and IGBTs. These components serve as the building blocks for controlling and converting electrical power in a highly efficient manner. By leveraging the fast-switching capabilities of these devices, power electronic systems can minimize energy losses while ensuring precise control over voltage and current levels. The ability to dynamically manage power

H. A. G. Al-kaf and K.-B. Lee, *Machine Learning Integration in Power Electronics*,
https://doi.org/10.1007/978-981-95-3845-4_1

flow makes these technologies crucial for diverse applications. Power electronics encompass several core conversion functions: AC to DC, DC to AC, DC to DC, and AC to AC. These conversion processes are the foundation for numerous applications, including renewable energy systems, electric vehicles, and industrial motor drives. For example, in solar power systems, inverters convert DC power from photovoltaic cells into AC power for grid compatibility, while DC-DC converters optimize battery performance in electric vehicles. The historical evolution of power electronics reflects its growing importance. Starting with the introduction of mercury arc rectifiers in the early twentieth century, the field has advanced through the development of silicon transistors and thyristors, leading to the current era of silicon carbide (SiC) devices and gallium nitride (GaN) devices. These innovations have drastically improved the efficiency, thermal performance, and power-handling capabilities of modern power electronic systems, enabling their use in high-temperature and high-frequency environments. Today, power electronics is a cornerstone of modern technology, enabling advancements in energy systems, transportation, and industrial automation. Its role in reducing energy consumption, enhancing grid stability, and supporting renewable energy adoption makes it an essential discipline for addressing global energy challenges. As power semiconductor technology continues to evolve, the applications and capabilities of power electronics will expand, driving progress toward a more sustainable and energy-efficient future.

1.2 Power Converters

Power converters are crucial in power processing and control systems for a wide range of applications, from small-scale devices like those used in energy harvesting to large-scale systems in power generation and transmission. These converters use semiconductor switches that operate at various frequencies, from 50/60 Hz to 100 MHz. Power electronics converters are classified based on several key characteristics as shown in Fig. 1.1.

1.2.1 Isolation

I. Nonisolated Converters: These are generally preferred when electrical isolation is not necessary because they are less bulky, cost-effective. They do not contain magnetic or electric isolation.
II. Isolated Converters: These converters typically use a transformer or coupled inductor to provide electrical isolation and to achieve purposes such as voltage level shifting, creating multiple outputs, or avoiding ground loops.

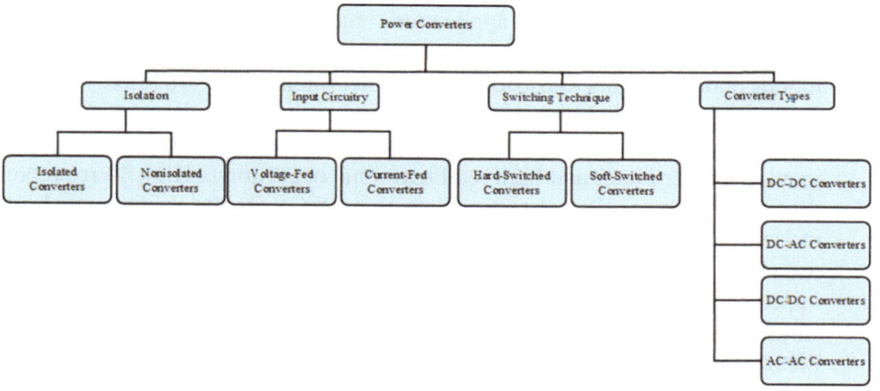

Fig. 1.1 General overview of classification of power converters

1.2.2 Input Circuitry

I. Voltage-Fed Converters: These are supplied by voltage sources with a capacitor connected in parallel with the source. The load for these converters acts such as a current source, and a filter inductor is typically connected in series with the load.

II. Current-Fed Converters: These are supplied by current sources, with an inductor connected in series with the source. They are commonly used in low-voltage renewable energy applications such as photovoltaics and fuel cells because their input inductors ensure a continuous, low-ripple input current.

1.2.3 Switching Technique

I. Hard-switched converters switch devices on and off while voltage or current is still present across them, causing high switching losses, electromagnetic interference (EMI), and stress on the devices.

II. Soft-switched converters, employing techniques like Zero Voltage Switching (ZVS) or Zero Current Switching (ZCS), reduce or eliminate voltage/current during transitions. This minimizes EMI and switching losses, enabling higher efficiency and longer lifespan. However, soft-switching requires more complex circuitry and control.

1.2.4 Types of Power Conversion

I. AC-DC Converters (Rectifiers) commonly referred to as rectifiers, are used to convert alternating current (AC) input into a direct current (DC) output. These converters are fundamental in applications requiring a stable DC voltage, which includes nearly all modern electronic devices such as televisions, computers, and mobile phone chargers. Basic forms of rectifiers include half-wave and full-wave diode rectifiers, which are cost-effective and simple to implement. However, their major drawback lies in the high current distortion they introduce, resulting in a low power factor and high total harmonic distortion (THD). To overcome these inefficiencies, Power factor correction (PFC) circuits are integrated into modern AC-DC converters. These circuits are designed to align the input current waveform with the voltage waveform, thereby reducing harmonic distortion and improving the overall power quality. With PFC, the input current becomes nearly sinusoidal and remains in phase with the input voltage, effectively reducing THD below the critical 5% threshold and enhancing energy efficiency.

II. DC-AC Converters (Inverters) also known as inverters, perform the reverse operation of rectifiers by converting a DC input voltage into an AC output voltage with controlled frequency and magnitude. This type of power conversion is essential for applications that require AC power from DC sources, such as renewable energy systems and motor drives. In renewable energy integration, inverters play a vital role in connecting solar photovoltaic (PV) or wind power systems to the AC grid. Similarly, they are widely used in variable-speed motor drives, which are crucial for industrial automation and energy-efficient motor operation. Inverters are generally classified into two main types: voltage source inverters (VSIs) and current source inverters (CSIs). VSIs control the output voltage, while CSIs regulate the output current. The type of waveform produced by the inverter also varies based on application requirements. Common output waveforms include square waves, modified square waves, and pure sine waves. Pure sine wave inverters are preferred for sensitive electronic devices due to their minimal harmonic content and compatibility with standard AC loads.

III. DC-DC Converters (Choppers) are designed to convert one level of DC voltage into another, either increasing (step-up) or decreasing (step-down) the voltage as required. These converters are integral to a wide range of battery-powered electronic devices, including smartphones, tablets, laptops, and other portable gadgets. One of the key applications of DC-DC converters is in energy harvesting systems, particularly those using renewable sources such as solar and wind energy. These converters optimize power extraction by adjusting the voltage levels to match load or storage requirements. DC-DC converters are broadly categorized into isolated and non-isolated types. Non-isolated converters are typically used when isolation between input and output is not required, offering higher efficiency and simpler design. An example of an advanced non-isolated topology is the dual SEPIC (Single-Ended Primary Inductance Converter) PWM switching-mode DC-DC converter. This type offers both step-up and step-down capabilities with improved ripple performance and voltage regulation.

IV. AC-AC Converters are used to directly convert one form of AC power (specific voltage and frequency) to another. This conversion is crucial in applications requiring voltage level changes or frequency adjustments without an intermediate DC stage. Such applications include variable-speed motor drives, HVAC systems, and industrial automation processes. These converters can be broadly divided into direct and indirect types. Indirect converters typically use a DC-link stage consisting of a rectifier followed by an inverter, while direct converters, such as cycloconverters and matrix converters, perform conversion without energy storage components. Cycloconverters are naturally commutated and are suitable for low-frequency applications such as large motor drives. Matrix converters, on the other hand, use fully controlled bidirectional switches, offering compactness, bidirectional power flow, and sinusoidal input/output waveforms.

1.3 Overview of AC Motor Drives

The role of electrical alternating-current (AC) drive machines in converting electrical energy into mechanical energy has been pivotal across various industrial applications over recent decades. This includes pivotal sectors such as electric vehicles (EVs), modern servo systems, robotics, chip-mount machines, and belt conveyors. Figure 1.2 provides a general overview of an advanced drive system for AC motor drives, highlighting the key factors that have driven these advancements. Firstly, improvements in power electronic devices have played a key role in advancing AC motor drives. These improvements include achieving higher power density, increasing switching frequencies, reducing power losses, expanding power capacity, and improving cost-effectiveness. These changes have fundamentally transformed the landscape, creating more efficient and reliable AC motor drives. Additionally, progress in mechanical structure and material technologies has been crucial in enhancing the performance and efficiency of AC motor drives. Innovations in superconducting materials, permanent magnet materials, and new structural materials have revolutionized motor design and operation, resulting in increased efficiency and power output.

1.3.1 Classification of AC Motor

In recent years, AC motors, including switched-reluctance motors (SRMs), induction motors (IMs), and PMSM, have been widely utilized in various applications. Figure 1.3 illustrates the basic configuration of an AC motor assembly, which involves several key components along with various ancillary parts that support its operation. Among these, induction motor drives are particularly dominant in the industry due to their simplicity, low cost, durability, and minimal maintenance requirements. Figure 1.4 provides a classification of common synchronous motors and highlights the increasing prominence of PMSM. This rise in popularity can be attributed to their superior power density and efficiency compared to other AC machines, along with

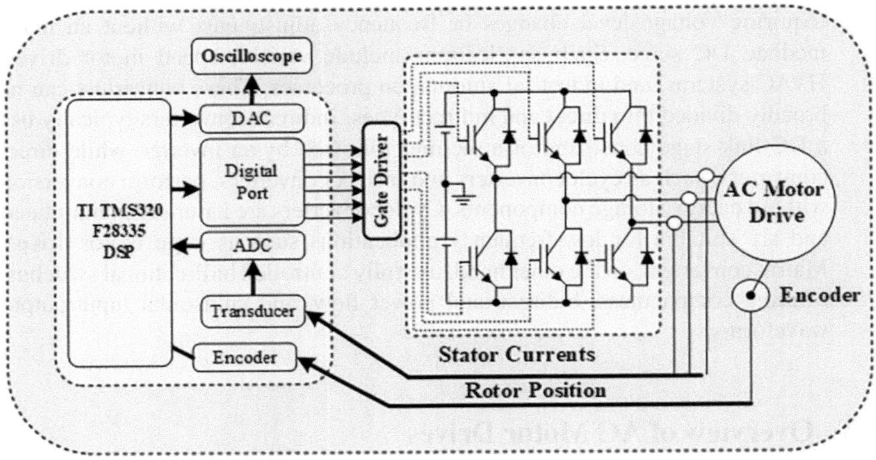

Fig. 1.2 General overview of advanced AC drive system

the growing emphasis on energy efficiency in response to escalating fossil fuel costs and climate change concerns. As a result, PMSMs have been extensively employed across sectors such as electric vehicles, industrial machinery, aerospace, domestic appliances, and robotics, where their compact size, wide speed range, high torque to power ratio, and operational efficiency offer significant advantages. As seen in Fig. 1.3, PMSMs can be configured in two ways: mounted on the surface of the rotor (SPMSM) or buried inside the rotor (IPMSM). In SPMSMs, the d-axis and q-axis inductances ($L_d = L_q$) are identical due to the rotor geometry, enabling torque generation through the interaction between the q-axis component of the stator currents and the magnetic flux. Conversely, in IPMSMs, the stator inductances differ ($L_d \neq L_q$) because of the non-uniform structure of the permanent magnet (PM) in the rotor. Accordingly, the induced torque is a combination of reluctance torque derived from the difference between the stator inductances and the stator currents and magnetic torque.

1.3.2 Control Classification of AC Motor

The control design of AC systems presents challenges due to the need for precise control output under non-linear conditions, uncertainties, and unmodeled parameters in power electronics converters. Research has focused on developing effective control systems for AC engines, leveraging advancements in power electronics and digital signal processors (DSPs) since the 1960s. Figure 1.5 outlines control techniques for AC motor drives, such as scalar V/f control, vector control, and predictive methods. FOC is a widely used technique that regulates torque and flux by controlling stator currents (d-axis and q-axis). DTC directly manages torque and flux for a

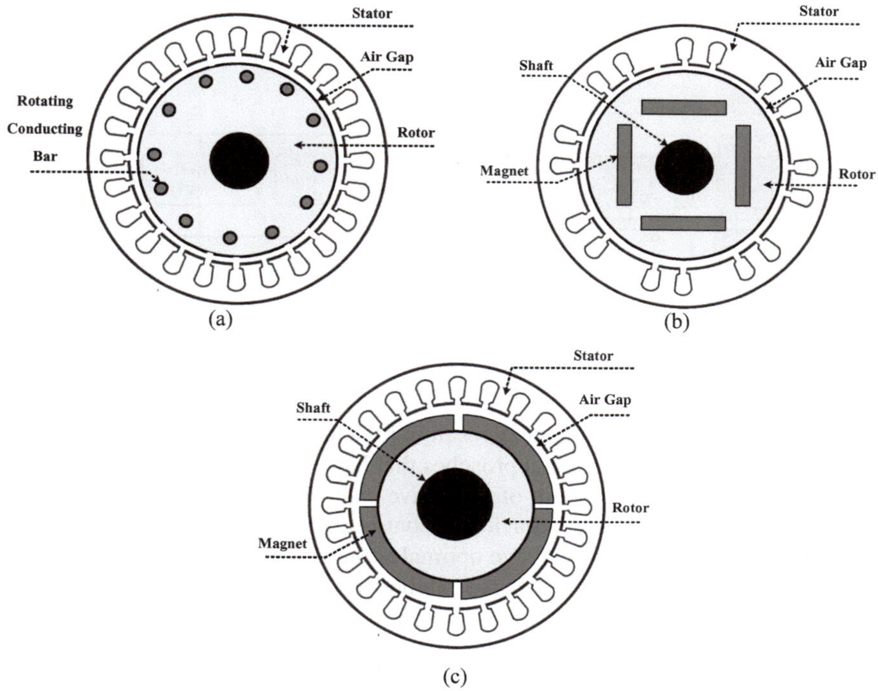

Fig. 1.3 AC motors in diverse sectors in industries. **a** IM. **b** IPMSM. **c** SPMSM

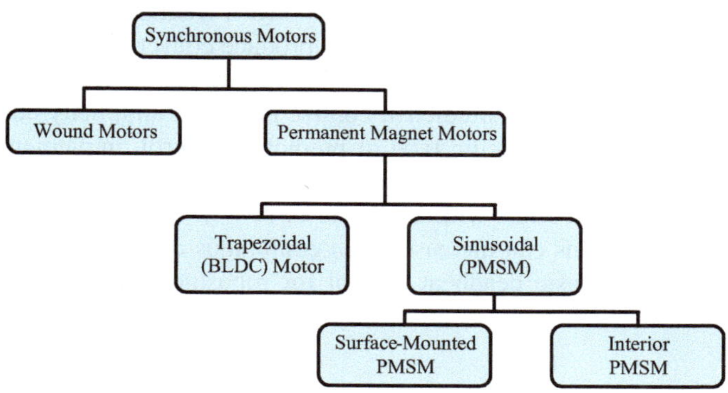

Fig. 1.4 Classification of synchronous AC motors according to their permanent magnet

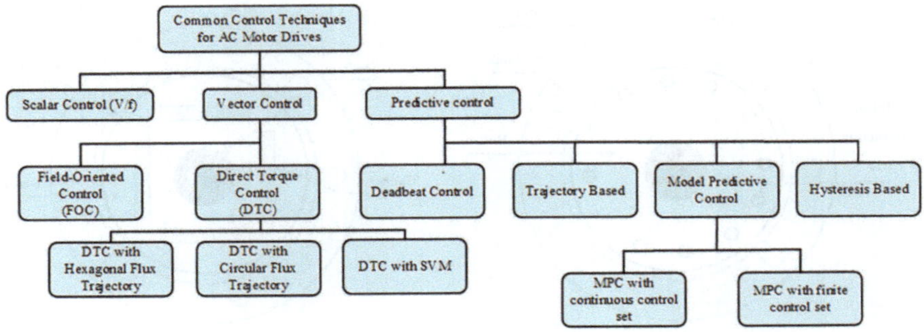

Fig. 1.5 Common classification of control techniques for AC motor drives

rapid dynamic response, which may result in high torque ripples. Predictive control methods offer a broad range of approaches that have seen recent adoption in power converters. One defining feature of predictive control is its utilization of a system model to forecast the future behavior of controlled variables. The controller then employs this information to achieve optimal actuation based on a predefined optimization criterion. In hysteresis-based predictive control, the optimization goal is to maintain the controlled variable within specific hysteresis bounds. Conversely, trajectory-based control mandates that variables follow predetermined trajectories. Deadbeat control aims to make the error zero in the next sampling instance. MPC employs a more adaptable approach expressed through a cost function to minimize. Among these methods, deadbeat control and MPC with continuous control sets necessitate a modulator to generate required voltages, leading to a fixed switching frequency. Conversely, other methods directly produce converter switching signals, avoiding the need for a modulator and resulting in variable switching frequencies. One of the key advantages of predictive control lies in its simplicity and intuitive nature. Depending on the specific type of predictive control, implementation can be straightforward, as seen in deadbeat control and finite control set MPC, particularly for two-level converters with a horizon of $N = 1$. However, the complexity of MPC implementations can increase when continuous control sets are involved. Similarly, enhancing basic deadbeat control for robustness can lead to complex implementations.

1.4 The Role of Machine Learning in Power Electronics

In recent years, there has been growing interest in applying ML techniques to power electronic systems [1]. This surge in interest is largely driven by the rapid advancements in ML, which have revolutionized both the algorithmic approaches and computational capabilities of power converter controllers. These improvements have enabled ML systems to address the complexity inherent in power electronics

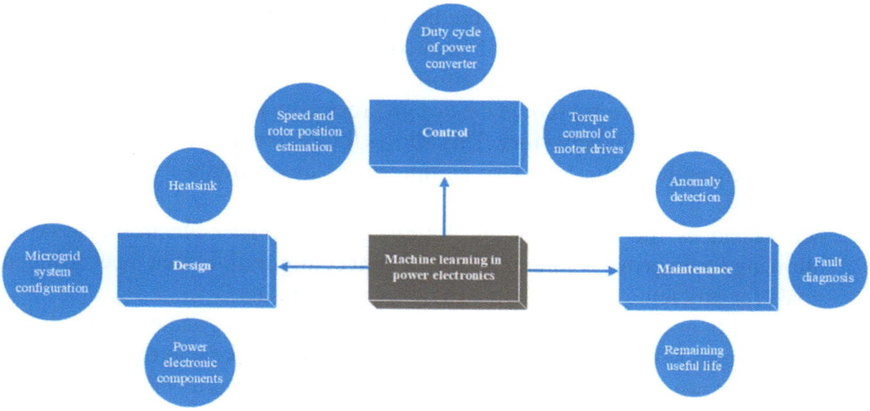

Fig. 1.6 Block diagram of machine learning applications in power electronics

more effectively due to their ability to model complex, nonlinear, and time-varying system behaviors that traditional control methods often struggle to handle. ML provides a flexible approach that learns from data, enabling systems to generalize to novel scenarios and adapt to changing operating conditions. This makes it ideal for dynamic environments where system parameters and external conditions vary significantly, such as in control systems. Furthermore, ML helps reduce computational complexity by approximating complex functions, which is crucial for real-time and embedded applications. It enhances system performance by optimizing control strategies, leading to improved efficiency, reduced energy losses, and higher output quality. Additionally, the robustness of ML methods allows systems to remain functional under variations or disturbances, boosting fault tolerance and system reliability. ML is crucial in several key areas of power converter-based systems, namely design optimization, intelligent control, and predictive maintenance as shown in Fig. 1.6 [2–4].

1.4.1 ML for Design Optimization

Designing power converter systems involves evaluating numerous potential configurations to arrive at the optimal solution, a process that demands considerable computational resources. Each candidate design must be simulated across various operational scenarios. This complexity is heightened by the need to balance multiple performance metrics, such as efficiency, power density, cost, and reliability, which can be interdependent. ML plays a crucial role in addressing the following challenges:

I. ML algorithms intelligently narrow down the design space by focusing on the most promising configurations, thereby reducing the number of simulations needed to find optimal solutions.

II. ML models can be trained to predict the performance of various designs, enabling faster assessments and iterative improvements of power converters without extensive computational effort.
III. ML serves as a powerful tool in the design process by creating surrogate models that replicate system behaviors. This allows for quicker evaluations of complex objectives without the intensive computations associated with direct system modeling.
IV. ML enhances the modeling and optimization of power electronic systems, allowing for improved circuit and component configurations that meet stringent goals for efficiency, power density, and size constraints.
V. Optimization tasks, such as heatsink design or microgrid system configuration, benefit from ML-driven approaches that refine design parameters to achieve a balance between size, temperature control, and overall performance.

1.4.2 ML for Control

ML has become increasingly significant in the control and monitoring of electric machine drives, particularly for replacing conventional PI and PID controllers [5]. These traditional controllers are widely used in industry due to their simplicity and low computational cost. However, they often struggle to achieve optimal performance when the controlled system is nonlinear and uncertain. ML offers advanced capabilities to model and control these systems more effectively by learning relationships between inputs and outputs, even when mathematical models are unavailable or overly complex. ML can be applied in various control scenarios, including real-time decision-making, adaptive control, and dynamic system optimization. For example, reinforcement learning can optimize power converters in real time by adapting to load and environmental changes. Supervised learning-based models can predict system behavior and enable efficient trajectory tracking and control signal generation. One of the critical benefits of ML in control is its ability to generalize and adapt. Neural networks, for instance, can be trained to approximate control laws, providing robust solutions under varying conditions. ML-based controllers can also integrate with traditional methods, enhancing system stability and performance. These hybrid approaches ensure reliability in systems such as renewable energy inverters and motor drives. ML plays a crucial role in addressing the following challenges:

I. ML improves the accuracy of speed and rotor position estimates, effectively addressing issues associated with parameter changes and nonlinearities.
II. ML provides precise temperature estimates to help manage motor health, preventing overheating and extending component longevity.
III. ML is increasingly used to enhance the performance of power electronic converters under dynamic environmental conditions, improving efficiency in wind and photovoltaic (PV) systems.

A. ML enhances energy capture by improving tracking and adjusting turbine rotational speed to optimize power output.
B. ML is leveraged to stabilize tracking under intermittent and fluctuating solar irradiance, maintaining efficient power output even during varying conditions.

IV. ML strategies are widely adopted in DC/DC converters to optimize the duty cycle directly, improving stability and performance in various applications.

1.4.3 ML for Maintenance

ML has demonstrated significant promise in the field of predictive maintenance for power converter systems. Traditionally, maintenance schedules rely on fixed intervals or post-failure repairs, leading to inefficiencies and higher costs. In contrast, ML-driven predictive maintenance utilizes real-time data from condition monitoring systems to forecast potential component failures, allowing for proactive maintenance before issues arise. ML applications in maintenance include anomaly detection, fault diagnosis, and reaming useful life (RUL) prediction [6, 7]. For instance, supervised learning models can classify operational states of components and identifying early signs of failure. Similarly, regression models can predict how long a component will function under given conditions, allowing for timely interventions. The integration of ML in maintenance provides unparalleled precision in condition monitoring. By processing large volumes of sensor data, ML algorithms can identify subtle trends and patterns that indicate degradation. This is particularly useful for critical components like capacitors, IGBTs, and power modules, where traditional monitoring methods may fall short. ML is crucial in fault detection for power electronic systems due to its ability to process complex, high-dimensional sensor data. Traditional methods often rely on predefined thresholds or manually designed rules, which may fail to detect subtle, early-stage faults. ML algorithms, especially supervised learning models, excel in recognizing patterns and anomalies in operational data, enabling the identification of incipient faults before they escalate. For instance, neural networks or support vector machines can process real-time data from sensors that monitor voltage, current, and temperature to detect deviations indicative of potential faults. ML techniques estimate critical parameters, such as capacitance, impedance, and equivalent series resistance. By analyzing input–output relationships, ML infers component health non-invasively, facilitating effective monitoring without additional hardware.

References

1. Chen P, Zhao J, Liu K, Dong K (2023) A review on the applications of reinforcement learning control for power electronic converters. In: 2023 IEEE international conference on energy technologies for future grids (ETFG). IEEE, pp 1–6

2. Zhao S, Blaabjerg F, Wang H (2021) An overview of artificial intelligence applications for power electronics. IEEE Trans Power Electron 36(4):4633–4658
3. Khandelwal A, Kumar J (2019) Applications of AI for power electronics and drives systems: a review. In: Proceedings of innovations in power and advanced computing technologies (i-PACT). IEEE, pp 1–6
4. Zhang S, Wallscheid O, Porrmann M (2023) Machine learning for the control and monitoring of electric machine drives: advances and trends. IEEE Open J Ind Appl 4:188–214
5. Bose BK (2007) Neural network applications in power electronics and motor drives—an introduction and perspective. IEEE Trans Ind Electron 54(1):14–33
6. Liu R, Yang B, Zio E, Chen X (2018) Artificial intelligence for fault diagnosis of rotating machinery: a review. Mech Syst Signal Process 108:33–47
7. Zhang S, Zhang S, Wang B, Habetler TG (2020) Deep learning algorithms for bearing fault diagnostics—a comprehensive review. IEEE Access 8:29857–29881

Chapter 2
Machine Learning Fundamentals with Applications in Power Electronics

Abstract This chapter explores the integration of machine learning (ML) concepts within the realm of power electronics. It begins by discussing the foundational principles of ML, followed by an overview of its various types. Each type is explained using specific examples relevant to power electronics, illustrating their practical applications. Widely used ML algorithms in power electronics are examined, highlighting their advantages and providing simple explanations of their concepts and mathematical foundations. The general steps involved in the ML process are outlined, with real-time deployment considerations set aside for discussion in subsequent chapters. To enhance the understanding of these steps and the implementation of ML algorithms, a step-by-step guide is included to demonstrate their effectiveness in power electronics. Finally, this chapter presents a list of available resources, including datasets relevant to power electronics, as well as the software tools and programming languages necessary for building ML algorithms. This chapter aims to equip readers with a comprehensive understanding of the fundamentals of ML in the context of power electronics.

2.1 What Is Machine Learning?

ML is a branch of AI that enables machines to autonomously learn from data and past experiences by identifying patterns and making predictions with minimal human intervention. ML allows computers to operate without explicit programming by using algorithms to iteratively learn from data and improve over time. Through this process, systems can derive insights from large data sets and enhance their performance in response to new situations. In addition, deep learning (DL) is a sub-domain of ML that enhances this capability by training computers to mimic human learning from examples. Although it often requires more data, DL generally delivers superior performance compared to traditional ML algorithms.

H. A. G. Al-kaf and K.-B. Lee, *Machine Learning Integration in Power Electronics*,
https://doi.org/10.1007/978-981-95-3845-4_2

2.2 Types of Machine Learning

ML is broadly divided into three categories: supervised learning, unsupervised learning, and reinforcement learning. Figure 2.1 provides a overview of each category, including representative algorithms and their applications in power electronics.

2.2.1 *Supervised Learning*

With supervised learning, a machine is trained using a set of labeled data, where each element is composed of given input/output pairs. The machine learns the relationship between the input and the output. The goal is to predict behavior or make a decision based on previously given data. For example, the machine can be given the following input to obtain specific outcomes:

1. A set of integer numbers (current and voltage values) to estimate the speed of the motor.
2. A time series of data is transferred to images using a wavelet transform to estimate capacitor C [1].
3. A set of images (three-phase current) to classify the fault type in an inverter.
4. A set of integers (current, voltage, temperature, and irradiance) to classify faults in a photovoltaic (PV) panel.

In supervised learning, the model learns to map inputs to outputs by minimizing the errors between its predictions and the actual labels. After training, the model is evaluated on a separate test set to assess its performance and ensure that it generalizes well to new, unseen data. This approach is used for tasks such as classification and regression, where the goal is to make accurate predictions based on past examples.

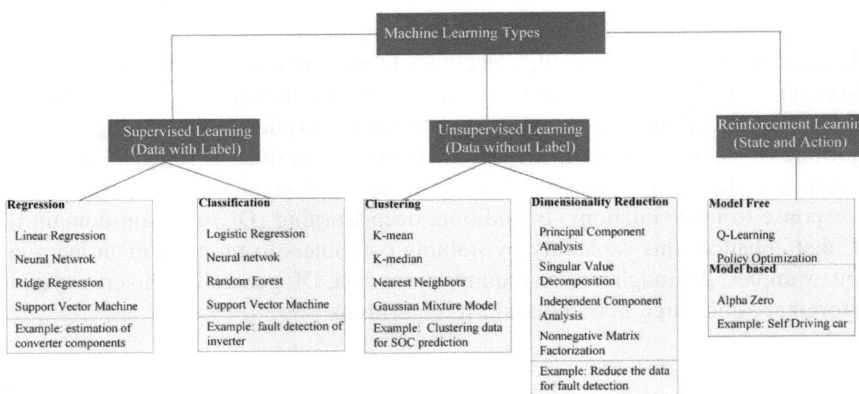

Fig. 2.1 Block diagram of machine learning types

2.2.2 Unsupervised Learning

In unsupervised learning, a machine is trained using unlabeled data, aiming to identify patterns or groupings based on the inherent characteristics or features of the data. These groupings are known as clusters. Unlike supervised learning, where specific answers are required, unsupervised learning evaluates the quality of the results based on how well elements within the same cluster exhibit similar characteristics or behaviors compared to those in other clusters. The effectiveness of this approach is measured by the degree of similarity among items within each cluster and the distinctions between different clusters.

Unsupervised learning encompasses several techniques, primarily focused on clustering and dimensionality reduction. Clustering involves grouping data points into clusters based on their similarities, whereas dimensionality reduction aims to decrease the number of features in a dataset while retaining important information.

To illustrate, several real-world examples of unsupervised learning applied in power electronics are presented:

1. Due to the lack of precise fault-condition experimental data, clustering methods can be utilized to recognize incipient fault patterns and assess the condition of a wound-rotor induction generator based on stator and rotor current and voltage signatures collected under both healthy and faulty conditions.
2. For anomaly detection in motor bearings, sensor signals are collected from the motor during normal operation without any labeled data. Then a clustering method is used to identify patterns representing the normal condition of the motor bearings. New signals that do not fit well into these clusters may indicate a potential failure [2].
3. Dimensionality reduction methods are used in fault detection to lower the dimensionality of data, which makes patterns and anomalies easier to identify. For instance, in fault diagnosis for a cascaded H-Bridge multilevel inverter system, voltage signals are first preprocessed using a fast Fourier transform. Then a reduction method is applied to extract key features and reduce data dimensions.

2.2.3 Reinforcement Learning (RL)

Reinforcement learning (RL) is a type of ML where an agent learns to make decisions by interacting with its environment. The agent takes actions and receives feedback in the form of rewards or punishments. Over time, the agent aims to maximize the cumulative reward by learning which actions lead to the most favorable outcomes. This learning process is driven by trial and error, allowing the agent to improve its decision-making based on past experiences and new information.

In the context of self-driving cars, RL can be used to train a vehicle to navigate and drive itself to a desired destination. The car, acting as the agent, gathers information about its surroundings through sensors and cameras. Based on these observations, it takes actions such as accelerating, braking, and steering. It receives rewards for actions that lead to safe and efficient driving, like staying in a lane and maintaining a safe distance from other vehicles. In addition, it receives penalties for actions that result in dangerous behaviors, such as sudden braking or collisions. Through trial and error, the car learns to maximize its cumulative reward, resulting in improved driving behavior over time.

In the context of controlling a permanent magnet synchronous motor (PMSM), RL can optimize the performance of the motor by replacing traditional PI controllers [3]. The RL agent interacts with a model of the motor, observing parameters such as the d-axis and q-axis currents and rotor speed. Based on these observations, the agent takes actions by adjusting the stator voltages to regulate the currents of the motor. It receives rewards for actions that help achieve desired speed and torque values while minimizing errors and control effort, and penalties for deviations or excessive control effort. Over time, the agent learns to maximize its rewards through trial and error, leading to improved motor control and performance.

2.3 Supervised Learning Algorithms

This section will introduce various supervised learning algorithms that are widely implemented in power electronics applications. Their concepts and mathematical models will be discussed to provide a guide on how these ML techniques learn and deliver high performance.

2.3.1 Artificial Neural Networks (ANNs)

ANNs are rapidly gaining traction in various power electronics applications. It is the most widely used algorithm in power electronics. ANNs have various applications, including enhancing the reliability of power converter components and improving control systems. For example, in motor drives, an ANNs have been implemented to improve the control performance of FOC by replacing PI control. Furthermore, ANNs reduce the computational burden of multi-step prediction in model predictive control (MPC) and select the optimal weighting factors for the MPC cost function [4]. Due to the power of ANNs to solve nonlinear problems, they are used in motor drives to estimate and handle nonlinearities and time-varying parameters. In addition, they are used to classify faults in motors, such as bearing faults, stator winding faults, rotor faults, and broken rotor bars, and to estimate motor parameters such as speed, torque, and position.

In this section, the components and structure of an ANN are explained, detailing how it learns and functions. The different layers, including input, hidden, and output layers are covered, and the roles of neurons, activation functions, weights, and biases are discussed. Additionally, the learning process of an ANN is explored, including

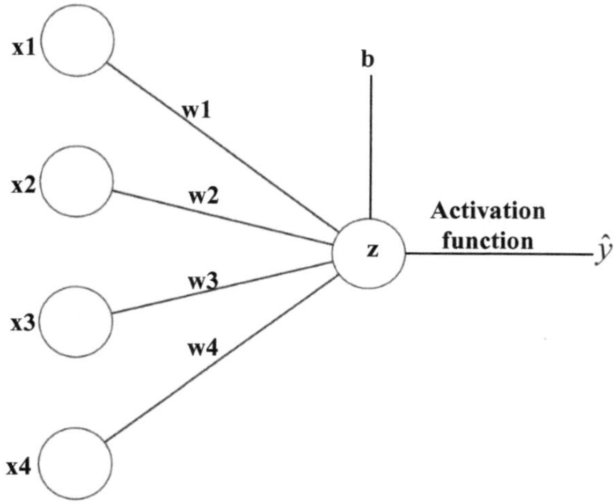

Fig. 2.2 Structure of a single layer neural network (SLNN)

forward propagation, loss calculation, and backpropagation with gradient descent, to provide a comprehensive understanding of how ANNs achieve effective learning and high performance. A single-layer neural network (SLNN), or "perceptron," is the simplest form of an ANN with just one output layer as shown in Fig. 2.2. This layer directly produces the final output based on input data that has passed through input neurons representing various features. There are three steps to implement SLNNs.

First, compute the weighted sum of the inputs. Multiply each input value x_i by its corresponding weight w and then sum these products as in (2.1). This weighted sum represents the total influence of the inputs on the output of the neuron. The weights determine how strongly each input affects the final result, with higher weights indicating a greater impact.

$$x \cdot w = (x_1 \times w_1 + x_2 \times w2 + \cdots + x_n \times w_n) \tag{2.1}$$

The inputs and weights are represented as vectors $x = [x_1, x_2, x_n]$ and $w = [w_1, w_2, w_n]$, respectively. The summation is calculated as the dot product of these vectors as presented in (2.2).

$$\sum_{i=1}^{n} x_i w_i \tag{2.2}$$

Next, add a bias term b to the weighted sum of the inputs. This adjusted sum is referred to as z as in (2.3). The bias term helps shift the activation function to the left or right, which is essential for fine-tuning the output of the network. By including the bias, the perceptron can better fit the data and make more accurate predictions.

$$z = x \times w + b \tag{2.3}$$

After computing the value of z, it is passed through an activation function to introduce non-linearity. This step is crucial because activation functions enable the network to model complex patterns and relationships in data, which purely linear models cannot capture. For perceptrons, a binary step function is commonly used. However, there are also several other activation functions, including sigmoid (or logistic) functions. The sigmoid function, as in (2.4), transforms the weighted sum plus bias into a value between 0 and 1, providing the final predicted output.

$$\hat{y} = \sigma(z) = \frac{1}{1 - e^{-z}} \tag{2.4}$$

The learning algorithm for a perceptron consists of backpropagation and optimization. Backpropagation, which stands for the backward propagation of errors, is used to compute the gradient of the loss function with respect to the weights. The loss function, such as the mean squared error for regression problems or the cross-entropy for classification problems, measures how far the predicted values are from the actual values. The average loss over the entire training dataset is referred to as the cost function (the mean squared error for regression, as presented in (2.5)).

$$C = MSE = \frac{1}{n}\sum_{i=1}^{n}(y_i - \hat{y}_i)^2 \tag{2.5}$$

To optimize the perceptron, it is crucial to understand how the cost function changes concerning the weights and bias. This involves calculating the gradients, which represent the rate of change of the cost function relative to the weights and bias. These gradients help in adjusting the weights and bias to minimize the cost function, thereby improving the performance of the perceptron. The gradient of the cost function C with respect to the weight w_i and bias b is computed using partial derivatives, as shown in (2.6).

$$\begin{aligned} \frac{\partial C}{\partial w_i} &= \frac{2}{n} \times sum(y - \hat{y}) \times \sigma(z) \times (1 - \sigma(z)) \times x_i \\ \frac{\partial C}{\partial b} &= \frac{2}{n} \times sum(y - \hat{y}) \times \sigma(z) \times (1 - \sigma(z)) \end{aligned} \tag{2.6}$$

Optimization involves choosing the best option from a set of possibilities. For a perceptron, this means finding the best weights and bias. One example of an optimization algorithm is gradient descent. Gradient descent adjusts the weights and bias based on the negative gradient of the cost function. The learning rate (α) is a hyperparameter that is used to control how much the weights and bias are changed. The process updates the weights and bias repeatedly until the model converges to the best solution, as shown in (2.7).

$$w_i = w_i - \left(\alpha \times \frac{\partial C}{\partial w_i}\right)$$
$$b = b - \left(\alpha \times \frac{\partial C}{\partial b}\right) \tag{2.7}$$

Due to their simplicity, single-layer neural networks can only handle linearly separable problems and are limited when it comes to representing complex data relationships. They are rarely used for real-world applications. Multi-layer perception (MLP) neural networks are much more capable of handling complex tasks.

The architecture of an MLP consist of three parts:

1. The input layer is the first layer in an MLP, consisting of neurons that pass data to subsequent layers without performing computations. The number of neurons matches the number of input features.
2. One or more hidden layers sit between the input and output layers. Each hidden layer consists of neurons that apply activation functions to their weighted inputs. The number of neurons in hidden layers can vary, and more layers can capture more complex patterns in data.
3. The final layer is the output layer, which generates the predictions of the network. The number of neurons depends on the problem type; one for binary classification and multiple for multi-class classification.

The training process of an MLP consists of two main phases: forward propagation and backpropagation, which are often optimized using gradient descent.

During forward propagation, input data is fed into the network through the input layer. Each neuron in the hidden layers computes a weighted sum of its inputs, applies an activation function (such as sigmoid), and passes the result to the next layer. This process continues until the output layer generates the predicted output. The predicted output is then compared to the actual target value, and the difference is calculated using a loss function, which quantifies the error.

Backpropagation follows forward propagation and involves updating the weights of the network based on the error calculated earlier. First, the gradient of the loss function with respect to the output is computed, indicating how the loss changes with respect to the output. This gradient is then propagated backward through the network using the chain rule of calculus, allowing for the calculation of gradients for each weight in the network. These gradients indicate how much each weight should be adjusted to reduce the loss.

Additionally, random initialization of the weights in the hidden layers is crucial. If all the weights are initialized identically, the neurons become indistinguishable, hindering effective training. Random initialization breaks this symmetry, enabling neurons to learn diverse features.

Activation functions are essential in MLPs because they introduce non-linearity, enabling the network to model complex functions. Without them, an MLP would act as a SLNN, regardless of the number of layers. Different types of activation functions exist, including hyperbolic tangents (Tanh), and rectified linear units (ReLU), as shown in Eqs. 2.8.

$$\tanh(z) = 2\sigma(2z) - 1$$
$$ReLU(z) = \max(0, z) \tag{2.8}$$

The tanh function is similar but it outputs values between –1 and 1, which can speed up training. The ReLU function, which is popular in deep learning, outputs the input if it is positive and zero if it is negative. Despite not being differentiable at zero and having a zero gradient for negative inputs, ReLU is effective in practice and helps mitigate the vanishing gradient problem. The logistic (sigmoid) function maps real numbers to values between 0 and 1, making it useful for backpropagation due to its S-shape and non-zero derivative. Activation functions are crucial because they allow MLPs to learn from errors and adjust their weights, enabling the network to model intricate data relationships. Without them, multiple linear transformations result in a linear transformation, making it impossible to solve complex problems. For example, choosing two linear functions still yields a linear function, so non-linearity is necessary for depth to add complexity. In summary, the backpropagation algorithm for each training instance involves a forward pass to compute the output of the network, error calculation using a loss function, a backward pass to compute the gradient of the loss with respect to each weight using the chain rule, and a gradient descent step to update the weights and minimize the error. This process is repeated for each mini-batch of training instances, and the network is trained over multiple epochs until it converges to a solution.

Figure 2.3 shows a regression or prediction problem, where an MLP is used to estimate the speed of a PMSM. The process begins by collecting data from the PMSM speed control system, where the inputs to MLP are the direct axis current (I_{ds}), quadrature axis current (I_{qs}), direct axis voltage (V_{ds}), and quadrature axis voltage (V_{qs}) in the stationary reference frame. The target output for the MLP is the actual motor speed (ω_r). The data collected covers a wide range of operating conditions, including low, medium, and high speeds with varying voltages and currents. This diverse dataset allows the neural network to learn and predict the motor speed accurately across different operational scenarios. The Levenberg–Marquardt algorithm is used to train the network, optimizing it to minimize the error between predicted and actual speeds.

Figure 2.4 shows a classification problem, where an MLP is used to emulate the performance of an MPC system. The inputs to this network include the direct axis current (I_{ds}), quadrature axis current (I_{qs}), and their reference values, while the target output is derived from the optimal voltage vectors predicted by the MPC model. The role of the MPC controller is to predict and optimize the behavior of the system over a future time horizon based on the current state and reference inputs. By training the MLP to replicate the control actions and predictions of the MPC system, this approach allows for an approximation of the performance of the MPC without the computational complexity typically associated with real-time optimization. The network learns to predict the optimal control inputs required to achieve the desired system performance, thereby providing a streamlined alternative to the full MPC approach.

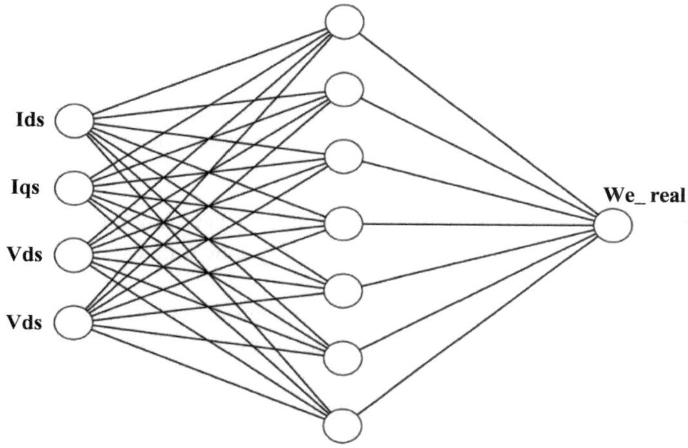

Fig. 2.3 Structure of a multi-layer perception neural network for estimating the speed of a PMSM

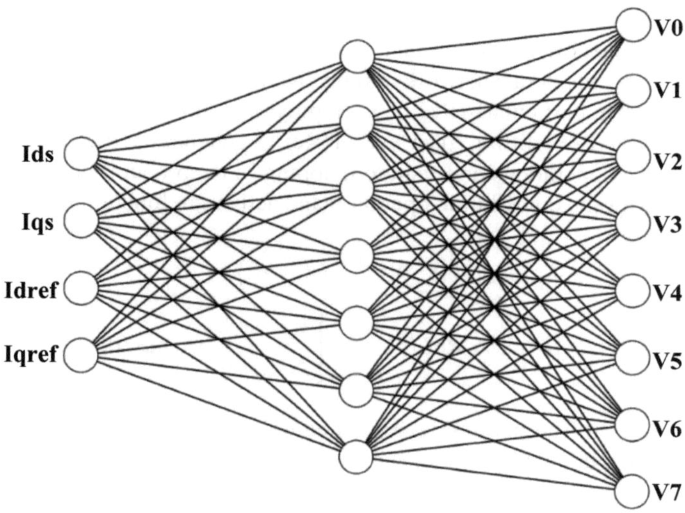

Fig. 2.4 Structure of an MLP for emulating the performance of an MPC method

2.3.2 Convolutional Neural Networks (CNNs)

CNNs are employed for feature extraction and fault classification to detect faults in various types of inverters [5]. In addition, they are used to estimate the power component of inverters.

The CNN consists of convolutional, pooling, and fully connected (FC) layers, as illustrated in Fig. 2.5. It has two key stages: feature extraction and classification prediction. In the feature extraction stage, deep features are learned layer by layer through the convolutional and pooling layers. The convolutional layers detect patterns in the input data, while the pooling layers reduce dimensionality and computational complexity. The classification prediction stage uses a softmax layer to classify and predict the learned features. Below is a detailed overview of the steps involved in a CNN.

(1) Input: The process begins with the input layer, where raw image data is fed into the network. The input can be a 2D array (grayscale images) or a 3D array (color images with RGB channels).
(2) Convolutional layer: The first main component of a CNN is the convolutional layer. In this step, filters (or kernels) slide over the input image to perform convolution operations to detect specific features such as edges or textures. The output of this operation is called a feature map, which highlights the presence of these features in the input image. This layer accounts for the presence of repeated features that often appear at different positions, simplifying the FC neural network. It enables weight sharing among features and sparse connectivity for neurons, reducing the number of training parameters and preventing unnecessary repetitions during training. The convolutional layer effectively learns features irrespective of their positions, facilitating parallel operations that significantly accelerate network training. Nonlinear mapping through the activation function, particularly ReLU, is employed to mitigate issues like vanishing gradients and overfitting. The use of ReLU enhances the efficiency of both convolution operations and feature mapping in deep neural networks. Feature mapping for the convolution operation and activation function is given in (2.9).

$$\begin{aligned} X_i^l &= act\left(\sum X^{l-1} * w_i^l + b_i^l\right) \\ act(.) &= \mathrm{Re}LU(z) = \begin{cases} 0, \ z < 0 \\ z, \ z \geq 0 \end{cases} = \max(0, z) \end{aligned} \tag{2.9}$$

(3) Pooling layer: The pooling layer in a neural network re-extracts feature maps within a specific field to reduce dimensionality, accelerate training, and enhance

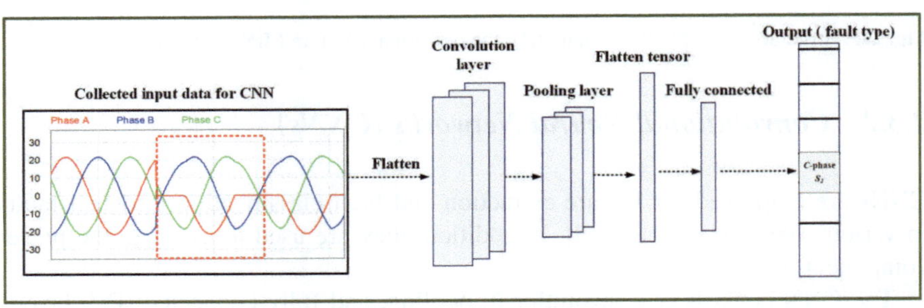

Fig. 2.5 Structure of a convolutional neural network (CNN)

feature robustness. It is a static attribute with no parameters that maintains the depth of the data while reducing the dimensions of each output. Pooling operations include maximum, average, and random types, and can be global or traditional in range.

(4) Fully connected layer: The fully connected (FC) layer is used as the intermediary between the final pooling layer and the classification layer. Typically, it is essential to transform feature maps from the last pooling layer into a one-dimensional array, which is accomplished through a flattening operation, before serving as the input for the FC layer. The FC layer can be mathematically expressed as (2.10).

$$FC(X) = \text{ReLU}(W \cdot X + B) \tag{2.10}$$

The softmax layer is a special part of the network that helps to classify and make predictions. It uses the softmax function for its operations. The output is expressed in (2.11).

$$\text{Output}(X) = \text{softmax}(W \cdot X + B) \tag{2.11}$$

The softmax layer connects the learned features to specific labels. It takes values from the previous layer of the neural network as the input and produces class scores as the output. The probability distribution for each class is represented using a normalized exponential function.

(5) Objective function: The cross-entropy loss function is preferred in various fault identification tasks. This function measures the difference between the true and predicted values. This indicates the difference between the actual output and the expected one. A lower cross-entropy value signifies a closer match between the true and predicted values. The goal of the objective function is to guide the training set of the network. Given a training set A $= \{x_n, y_n\}$ $\text{n}_{\text{N=1}}$, where y_n is the one-hot vector representation of the true label for the n_{th} data point x_n, and the final output of the network is denoted as $\hat{y}_n$, the objective function on the dataset A is determined in (2.12).

$$Loss = -\frac{1}{N}\sum_{n=1}^{N} y_n^T \log \hat{y}_n \tag{2.12}$$

The network parameters were adjusted by minimizing the objective function. A smaller loss implies a closer match between and y_n, further indicating that the predicted value approached the true value.

CNN is used as a classification algorithm to predict the faults in the NPC inverter, while processing normalized three-phase current data. The normalization step divides the current data by the respective current level for each simulation and experiment, resulting in a 3 × N matrix, where N represents the number of current data points in one cycle. This preprocessed data serves as the input for the CNN, which is designed to detect 12 different open-switch fault cases in a three-phase circuit with four switches (S1 to S4), alongside one label for normal operation, resulting in a total of 13 labels. The CNN architecture consists of convolutional layers with 3×1 kernels, followed by max pooling layers that reduce dimensionality and emphasize significant features. These convolutional layers capture local patterns and extract complex features from the input data. Subsequently, a fully connected layer aggregates these features, and dropout is applied to prevent overfitting.

2.3.3 Support Vector Machines (SVMs)

SVMs are supervised learning algorithms that have been successfully applied to the problems of classification and regression. In SVMs, the goal is to find a decision boundary, or hyperplane, that effectively separates data points belonging to different classes. The optimal hyperplane, as shown in Fig. 2.6, is the one that maximizes the margin, which is the distance between the hyperplane and the closest data points from each class. These data points, known as support vectors, are critical because they directly influence the position and orientation of the hyperplane, thereby defining the classification boundary. By maximizing the margin, SVMs aim to improve the classifier robustness, ensuring that it can generalize well to unseen data. The process of finding the optimal hyperplane involves generating various hyperplanes that can potentially separate the classes. Among these, the best hyperplane is selected as the one that offers the largest margin between the support vectors of the two classes. This hyperplane is known as the maximum margin hyperplane. The wider the margin, the more confidence the model has in its classification, reducing the risk of misclassification on new data. When data points are too dispersed to be separated by a linear hyperplane, SVMs employ a kernel trick to map the input space to a higher-dimensional space where linear separation becomes possible. For linearly separable data, a linear kernel is often used, especially in scenarios with a large number of features. This kernel is favored because its training is faster, requiring optimization of only the C regularization parameter. In contrast, other kernels also require optimizing the γ parameter, making grid search more time-consuming. They are used in a variety of applications, including predicting IGBT junction temperature, estimating the battery state of charge, and classifying the faults in inverters [6].

The mathematical method of the linear kernel of ε-SVM is implemented to emulate and outperform the performance of MPC for power converter, as shown in Fig. 2.7 [7].

The linear kernel function in an SVM is the simplest kernel and it is defined as the dot product of two vectors, as presented in (2.13).

$$\Phi(x_j, x_k) = \langle x_j, x_k \rangle = x_j^T \cdot x_k \tag{2.13}$$

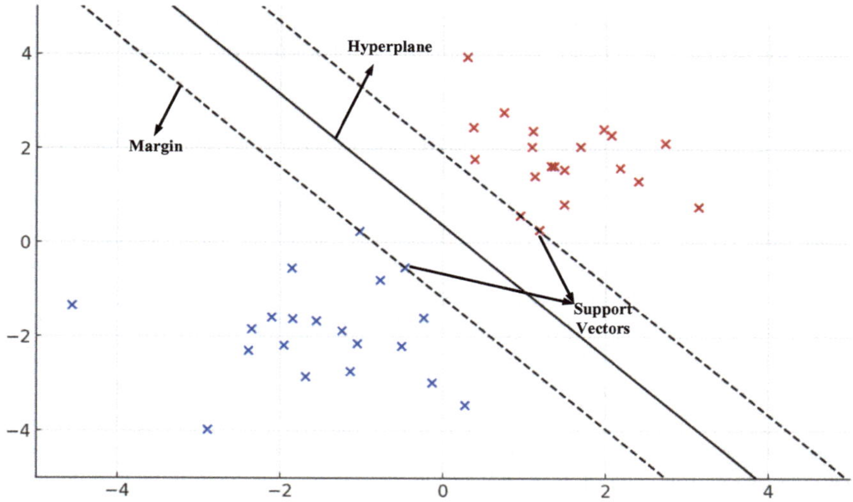

Fig. 2.6 Optimal hyperplane of a support vector machine (SVM)

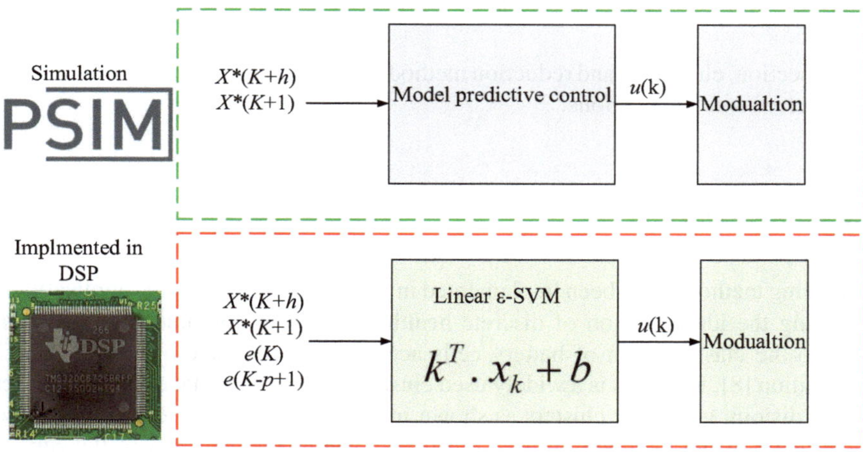

Fig. 2.7 Linear kernel of ε-SVM for emulating the performance of an MPC

where x_j and x_k are two vectors in the input space, and $x_j^T \cdot x_k$ denotes the dot product, which measures the similarity between x_j and x_k.

Then the prediction function of the ε-SVM with a linear kernel is given in (2.14):

$$\tilde{u}_k = \sum_{j=1}^{m} \beta_j \cdot x_{sv,j}^T \cdot x_k + b = k^T \cdot x_k + b \tag{2.14}$$

where $x_{sv,j}$ are the support vectors, which are data points that lie on the margin or support the decision boundary, x_k is the input vector for making a prediction, b is the bias, which helps in shifting the decision boundary, and β represents the weights for the support vectors.

The ε-SVM can emulate and potentially outperform traditional online MPC by learning from the current and future states used by MPC and by incorporating historical errors. While MPC typically makes decisions based on inputs like the current measurement state variables and future references, it lacks the ability to capture trends from past information. In contrast, the ε-SVM–MPC model uses a more comprehensive input set, including historical errors along with the current state variables and future references. The steps to emulate the MPC using ε-SVM are as follows. First, the simulation is carried out with MPC to generate the current state variables, future references, historical errors, and optimal switching signals or optimal voltage vectors. Then the generated data is used for training the linear ε-SVM offline. After training, K and b are optimized. In real-time, only K and b are implemented without the need for the computation time of MPC [7].

2.4 Unsupervised Learning Algorithms

In this section, clustering and reduction methods are presented in detail, using various power electronic applications.

2.4.1 K-Means

Clustering methods have been implemented in various power electronic applications, including the identification of discrete health states from continuous degradation data, to the classification of battery cells according to the state of charge (SOC) distribution [8]. K-means is a widely used clustering algorithm that divides a dataset into K disjoint groups or clusters as shown in Fig. 2.8. The main idea is to use the mean value of the members of each cluster, known as the centroid, as the central point that represents the cluster. This centroid is a vector in a space with arbitrary dimensions. K-means is a simple and intuitive method, making it a popular first choice for analyzing datasets with well-separated classes. However, the algorithm does not inherently know the optimal number of clusters. To determine a suitable number of clusters, the Elbow method is often employed. This method aims to minimize the sum of the squared distances between each sample and its corresponding centroid, a concept also referred to as the minimization of inertia. The k-means algorithm operates through a series of iterative steps. Initially, k random samples are selected as the starting centroids. The algorithm then calculates the distances between all the samples and these initial centroids to assign each sample to the nearest cluster. Once the clusters are formed, the centroids are recalculated based on the current members

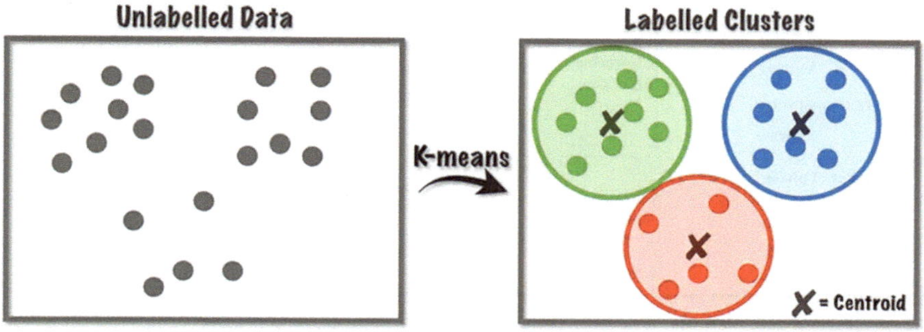

Fig. 2.8 Dataset division by the K-means method

of each cluster, and the process repeats. This results in changes in cluster membership as the centroids shift, prompting further recalculations. The process continues until a stop condition is met, such as when the centroids stabilize or after a set number of iterations.

The cluster assignment mechanism helps predict the SOC by grouping data into clusters based on features like voltage, current, temperature, and C-rate as shown in Fig. 2.9. Each cluster has an average set of these features, called a centroid. When a new data point is introduced, the model calculates the Euclidean distance between this point and the centroid of each cluster. The cluster with the smallest distance is considered the best match for the data point. This means the data point is assigned to that cluster, and the corresponding cluster model is used for the SOC estimation. This approach allows the model to adapt to different operational conditions, ensuring accurate and reliable SOC predictions. Cluster methods can also be implemented for fault detection as shown in Fig. 2.9. This process begins with the input of phase current data, where the averages and absolute values of these currents are calculated. This processed data is then fed into a clustering algorithm, which groups the data into different classes that represent normal and faulty operation modes. By analyzing these clusters, the system can detect deviations from normal conditions, identify potential faults, and label them accordingly. This method helps in accurately distinguishing between normal and faulty operations, facilitating effective fault detection and diagnosis.

2.4.2 Principal Component Analysis (PCA)

PCA is a technique used to reduce the dimensionality of large datasets by transforming data into a new set of variables, called principal components (PCs), as shown in Fig. 2.10. These PCs are uncorrelated and ordered by the amount of variance they capture from the original data. These components serve as a condensed representation of the data, capturing the most significant variance and patterns.

PCA begins with the standardization of data, especially when dealing with variables that have different units or scales. This step involves centering the data by

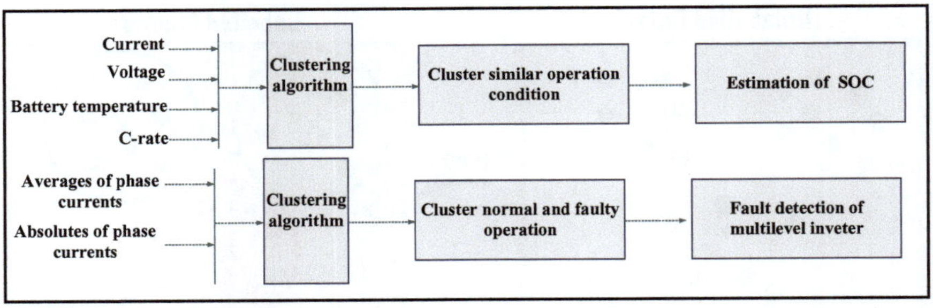

Fig. 2.9 K-means method for estimating SOC and detecting faults in a multilevel inverter

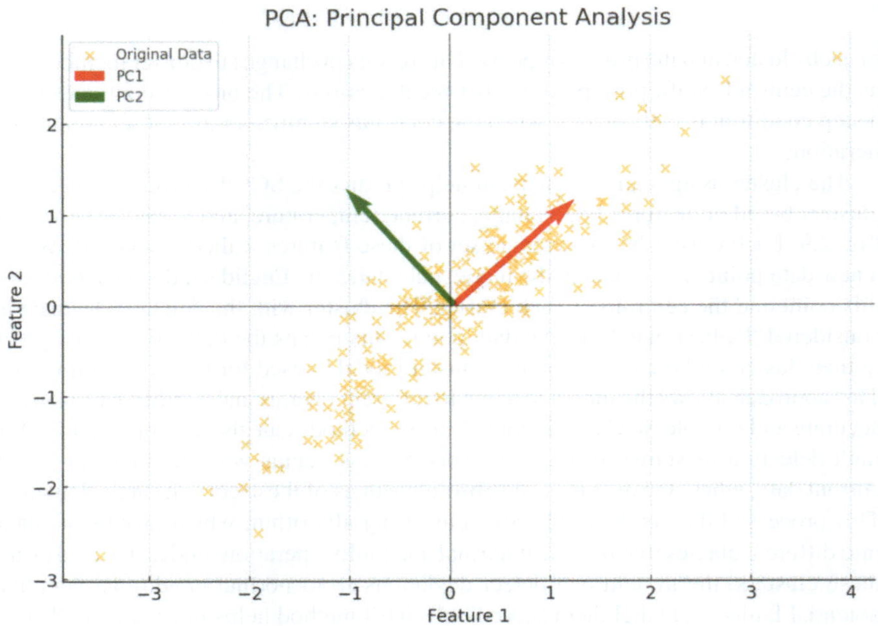

Fig. 2.10 Principal component analysis (PCA)

subtracting the mean from each value, then scaling it by dividing by the standard deviation. Standardization ensures that each variable contributes equally to the analysis and prevents variables with larger scales from dominating the results. Once the data is standardized, PCA computes the covariance matrix to understand how the variables relate to each other. It then calculates the eigenvectors and eigenvalues of the covariance matrix (or using singular value decomposition), which identify the principal components—directions in which the data varies the most—and quantify the variance captured by each component. The eigenvectors are then ranked by their eigenvalues, with the top eigenvectors becoming the principal components that capture the most variance. The original data is then projected onto these principal components, resulting in a simplified dataset with reduced dimensions that retains the most important information.

PCA is widely used in fault diagnosis to reduce the dimensionality of data, such as in detecting faults in induction motors [9]. PCA reduces input attributes such as voltage, current, rotational speed, and vibration. Each type of sensor data undergoes independent PCA, significantly reducing the number of input attributes while preserving essential information. The resulting principal components are then combined to form the input for the classification process.

2.5 Reinforcement Learning (RL) Algorithms

RL is a type of ML where an agent learns to make decisions by interacting with an environment. The agent receives rewards or penalties based on the actions it takes and learns to optimize behavior to maximize cumulative rewards over time. In RL, a state (s) represents the current situation of the environment, while an action (a) is the choice made by the agent to influence this environment. The reward (r) is the feedback received from the environment after an action is taken, indicating the success or failure of that action. The policy (π) is the strategy the agent follows to decide which actions to take based on the current state. Lastly, the value function (V) evaluates how advantageous it is to be in a particular state or to take a certain action within that state.

At time step t, the agent receives the current state s_t from the environment and selects an action a_t according to the policy π. In the subsequent time step $t + 1$, the environment responds by providing a reward r_{t+1} based on the action a_t and transitions to a new state s_{t+1}. The agent then updates the policy π based on the received reward r_{t+1}. The primary objective is to maximize the total cumulative reward.

2.5.1 *Q-Learning*

Q-Learning is a model-free algorithm used to learn the value of actions in different states without needing a model of the environment. It aims to find the optimal policy by learning Q-values, which represent the expected cumulative rewards of taking an action in a given state and then following the optimal policy.

The Bellman equation provides a way to update the Q-values based on the rewards received and the estimated future rewards. For Q-Learning, the Bellman equation is presented in (2.15).

$$Q(s, a) \leftarrow Q(s, a) + a[r + \gamma \max Q(s', a') - Q(s, a)] \tag{2.15}$$

where α represents the learning rate, which controls the extent to which new information updates old information. The reward r is the feedback received after taking action a in state s. The discount factor γ determines the significance of future rewards compared to the immediate rewards. The new state $s\prime$ is the state resulting from taking action a, and max $Q(s', a')$ denotes the highest Q-value for the new state $s\prime$, reflecting the maximum expected future reward.

The Q-learning algorithm operates through a series of well-defined steps to learn the optimal policy for an agent interacting with an environment. Initially, the Q-values, which represent the expected future rewards for state-action pairs, are commonly set to zero, though other arbitrary values can also be used. At each time step t, the agent observes the current state s_t and selects an action a_t based on a policy, such as the ε-greedy policy. The ε-greedy policy is a method for action selection that helps an agent explore the environment while also leveraging what it has already learned. It is designed to strike a balance between exploring new actions (which may lead to discovering better strategies) and exploiting the best-known actions (which maximize immediate rewards based on current knowledge). This policy involves choosing a random action with a probability of ε (exploration) and the action with the highest Q-value with a probability of $1-\varepsilon$ (exploitation). The agent then executes the chosen action, receives a reward r_t, and observes the new state s_{t+1}. The Q-value for the state-action pair (s_t, a_t) is updated using the Bellman equation, which incorporates the received reward and the maximum expected future reward from the new state. Over time, the agent transitions to the next state s_{t+1} and continues this process until the episode ends. The algorithm is repeated over many episodes to ensure that the Q-values converge to their optimal values. Key practical considerations include the learning rate α, which controls how new information influences the existing Q-values and typically decreases over time; the discount factor γ, which determines how future rewards are valued relative to immediate rewards; and the exploration rate ε, which balances the exploration of new actions with the exploitation of known rewards, and often decreases as the agent becomes more knowledgeable about the environment.

Q-learning is used to develop a direct torque control strategy through the Deep Q-Network (DQN) algorithm [10]. Deep Q-Learning is an advanced extension of the traditional Q-Learning algorithm that uses deep neural networks to handle complex and high-dimensional state spaces. Simulation results demonstrate that DQN control achieves effective dynamic performance, providing unified control under both constant torque and weak magnetic speed conditions without requiring detailed converter model parameters.

2.6 Machine Learning Workflow

In this section, a typical machine learning workflow is presented as shown in Fig. 2.11, which involves the following stages.

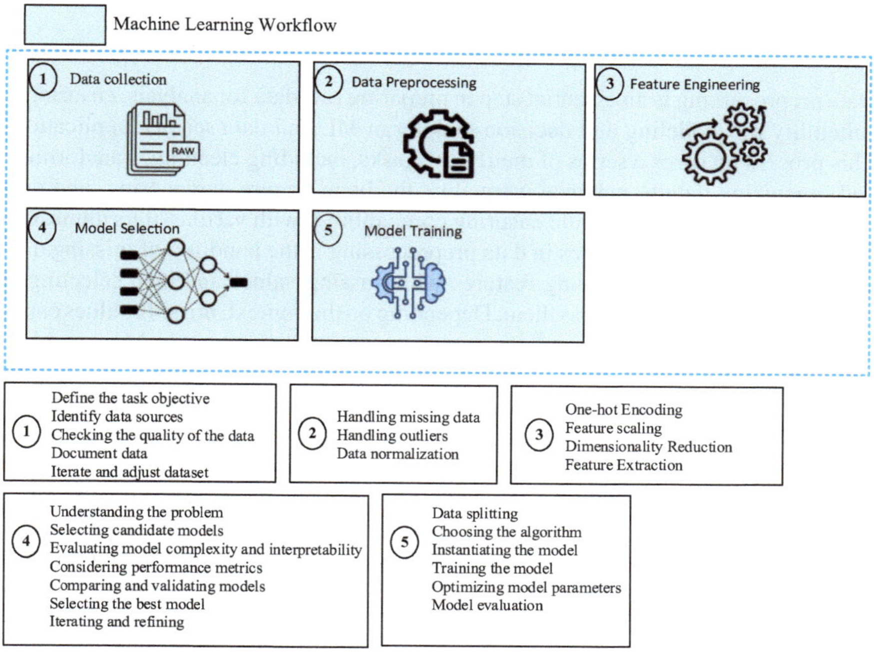

Fig. 2.11 Block diagram of the machine learning workflow

2.6.1 Data Collection

Collecting relevant data is a critical step in any ML project, since it directly influences the model performance and outcomes. The first step is to clearly define the task objectives, which helps guide efforts toward gathering the most relevant information. The next step is to identify potential data sources. Various open datasets related to power electronics are available on platforms such as Kaggle and IEEE DataPort, covering applications including motor datasets with thermal sensing in the rotor and stator, capacitor and ESR data for modular multilevel converters, renewable energy datasets, etc. After identifying potential data sources, it is crucial to rigorously assess the quality of the data by evaluating its accuracy and checking for errors, inconsistencies, or outliers that could affect the results. The completeness of the dataset needs to be assessed by confirming that it includes all the necessary variables and records needed to address project objectives. Consistency is another key factor. Data should be uniform across different sources and time periods, minimizing discrepancies that could complicate analysis.

2.6.2 Data Preprocessing

Data preprocessing is an essential step in preparing raw data for analysis, ensuring its suitability for modeling and decision-making in ML and data science applications. This process involves a series of meticulous tasks, including cleaning, transforming, and organizing data to enhance its quality, facilitate feature engineering, and optimize model performance while ensuring compatibility with various algorithms. One of the fundamental challenges in data preprocessing is the handling of missing data, which requires first identifying features with missing values and then selecting an appropriate strategy to address them. Depending on the context, missing values can be imputed using statistical measures such as the mean, median, or mode. More complex cases may call for predictive models to estimate the missing entries with greater accuracy. In situations where missing data is extensive or unreliable for imputation, an alternative approach is to eliminate the affected rows or columns entirely. For categorical data, a missing value can sometimes be represented by introducing a new category, or mode imputation can be applied to maintain consistency. Equally important is the management of outliers, which can significantly distort analytical results if left unaddressed. Identifying these anomalies often involves statistical methods such as box plots, Z-scores, or the interquartile range (IQR), with each method providing a structured way to detect extreme deviations from the norm. Once identified, outliers can be removed if they are deemed to be errors, capped by replacing them with the nearest non-outlier values, or transformed using techniques such as logarithmic, square root, or Box-Cox transformations to mitigate their impact. Consulting domain experts remains a valuable step in this process, since certain outliers may hold critical insights rather than being mere statistical anomalies. Beyond these considerations, data normalization plays a crucial role in ensuring that numerical features exist on a comparable scale, which is particularly important for machine learning models that are sensitive to varying magnitudes. Standardization, also known as Z-score normalization, min–max scaling, and robust scaling are widely used techniques to achieve this. Each of these techniques has advantages depending on the nature of the data. While standardization transforms data into a distribution with a mean of zero and a standard deviation of one, min–max scaling maps values into a fixed range, typically between zero and one. Robust scaling, on the other hand, leverages median and interquartile range to minimize sensitivity to outliers, making it particularly effective for datasets with extreme values. The preprocessing steps collectively serve as a foundation for robust and reliable analysis, ensuring that data is well-structured, informative, and optimized for ML models to derive meaningful insights.

2.6.3 Feature Engineering

Feature engineering is a critical process in ML that involves transforming raw data into a more structured and informative format to enhance its suitability for modeling. This step is essential for improving model accuracy, reducing overfitting, and ensuring that the model generalizes well to new data. Effective feature engineering encompasses creating new features, selecting the most relevant features, and transforming existing features to maximize predictive power. One of the most widely used techniques is one-hot encoding, which converts categorical variables into a numerical format by generating binary columns for each category. Dimensionality reduction techniques, such as PCA and t-SNE, can be applied to reduce the number of features while retaining essential information, thereby improving computational efficiency and model performance. Another important aspect of feature engineering is feature extraction, where new features are derived from existing features using mathematical transformations or domain-specific methodologies. This can include polynomial feature generation, which captures non-linear relationships, or statistical aggregations, which summarize key patterns within the data. By leveraging these strategies, feature engineering plays a foundational role in transforming raw data into a meaningful representation, ultimately driving more accurate and efficient ML models.

2.6.4 Model Selection

Selecting the right ML model for a specific task is a critical process that demands a careful assessment of multiple factors, including the nature of the problem, the characteristics of the dataset, the trade-offs between accuracy and interpretability, and the availability of computational resources. The process begins with a comprehensive understanding of the problem at hand, whether it involves classification, regression, clustering, or another form of ML task. This step requires a detailed examination of the data, including its features, target variables, size, distribution, and any inherent patterns or complexities that may influence model performance. Once the problem is clearly defined, domain expertise can guide the identification of potential models that are well-suited for similar tasks. These models may range from simple linear regression and decision trees to more sophisticated approaches such as support SVM, ANN, and ensemble methods, depending on the structure and complexity of the data. After narrowing down a selection of candidate models, it is crucial to evaluate their interpretability and computational demands. While deep neural networks and other highly complex models may deliver superior predictive accuracy, they often require significant computational power and present challenges in explainability. In contrast, simpler models such as decision trees and logistic regression offer greater transparency, which can be particularly valuable in applications where interpretability is a priority. Performance metrics must also be chosen with precision, ensuring that they align with the specific task. For classification problems, metrics such as accuracy, precision, recall, and F1-score provide valuable insights, while regression tasks are typically assessed using mean squared error (MSE) and R-squared values. Once the models are evaluated, fine-tuning their hyperparameters becomes essential to optimizing performance. Techniques like grid search, random search, Optuna, and Bayesian optimization can refine model parameters to achieve better results while balancing complexity and efficiency. The process of model selection is rarely linear.

Instead, it is often an iterative cycle that involves repeated refinements in feature engineering, hyperparameter tuning, and algorithm selection until the optimal balance is achieved. A well-chosen model maximizes predictive power and ensures robustness and generalizability to unseen data, making it a cornerstone of effective ML applications.

2.6.5 Model Training

Training a ML model is a structured process that involves fitting the selected algorithm to the available data to uncover underlying patterns and relationships. This process begins with splitting the dataset into training and validation sets, typically following a 70–30 or 80–20 ratio, to ensure the validation set accurately represents the real-world data distribution. By reserving a portion of the data for validation, the ability of the model to generalize beyond the training set can be effectively assessed. Once the data is prepared, the choice of an appropriate algorithm depends on the nature of the problem, whether it is classification, regression, or clustering, as well as the characteristics of the dataset. After selecting the model, it must be instantiated and initialized with appropriate parameters. For example, in Python's Scikit-Learn library, a logistic regression model can be initialized with a simple line of code, such as: `model = LogisticRegression()`. The next step involves training the model by fitting it to the training data using the `.fit()` method, enabling it to learn from the provided features and adjust its internal parameters accordingly. Once the initial training phase is complete, the focus shifts toward optimizing the model performance through hyperparameter tuning. Various techniques, including grid search, random search, and Bayesian optimization, can be employed to refine the hyperparameters and enhance the predictive capabilities of the model. After tuning, the model is evaluated on the validation set using appropriate performance metrics such as accuracy, precision, recall, or mean squared error, depending on the specific machine learning task. If the model achieves the desired level of performance, a final retraining step is conducted using the entire dataset, including the validation data, to maximize learning and prepare the model for real-world deployment. This final step ensures that the model has learned from the entire range of available data, enhancing its robustness and accuracy before being implemented in practical applications.

2.6.6 Workflow for Predicting Core Loss in Magnetic Materials

This section outlines the ML workflow developed to predict the average volumetric core loss (P_v) of power magnetic materials using data from the MagNet Challenge 2023, which considers widely known datasets in power electronics and aims to advance ML techniques for electromagnetic modeling. The motivation for applying

ML to magnetic core loss modeling arises from the limitations of traditional empirical equations such as the Steinmetz equation (SE) and its variants. While these equations offer simplicity, they often lack accuracy when capturing the complex behaviors of magnetic materials across diverse operating conditions, such as varying frequencies, waveforms, temperatures, and flux DC bias. Additionally, the SE framework requires manual curve-fitting and is limited by a small number of parameters, which restricts its predictive capabilities, particularly for new materials. ML methods provide a data-driven approach to overcome these limitations by learning intricate relationships directly from large datasets, improving accuracy, versatility, and scalability. Key considerations when applying ML include achieving high prediction accuracy with minimal parameters, ensuring computational efficiency for practical design tools, ensuring robustness across a wide range of operating conditions, and enhancing physical interpretability to improve the understanding of magnetic material behaviors.

The MagNet Challenge dataset provides systematically measured data for power magnetic materials operating in steady-state conditions [11]. Each data point consists of three key inputs: a single-cycle flux density waveform $B(t)$ sampled in 1024 steps, an operating frequency f_{sw} in hertz, and a corresponding temperature T in degrees Celsius. The dataset captures material behavior under different excitation waveforms, including sinusoidal, triangular, and trapezoidal shapes, to reflect real-world power converter conditions. The corresponding output is the volumetric core loss Pv (in W/m^3), which is determined by the area of the *B-H* loop. The dataset was strategically divided into training and testing subsets across different materials. The dataset includes materials such as 3C90, 3C94, 3E6, 3F4, 77, 78, N27, N30, N49, N87, 3C92, T37, 3C95, 79, and ML95S, covering a range of flux densities, frequencies, and temperatures. The dataset allows for a comprehensive evaluation of predictive models. This diversity enables models to generalize effectively across different materials and operating conditions while balancing accuracy and computational efficiency.

The data preprocessing involves analyzing and extracting useful information from the raw *B-H* loop measurements. The area enclosed by the *B-H* loop, which determines the core loss, is calculated using numerical integration techniques such as the trapezoidal rule. This step is particularly important when handling nonsinusoidal waveforms like triangular or trapezoidal shapes, as integration accuracy can be affected by sharp transitions or non-smooth curves. Given the nonlinear behavior of materials under varying conditions, care is taken to process the waveforms effectively. Additionally, features such as operating frequency and temperature are standardized to ensure consistency across the dataset. This step helps improve the performance of predictive models when generalizing across unseen operating conditions.

Data collection, preprocessing, feature engineering, model selection, and training will be discussed with Python code snippets based on the scikit-learn library. The training dataset comprises measured *B-H* loops, operating frequency, and temperature conditions.

In this demo, the feedforward neural network (FNN) is synthesized using the PyTorch framework. The first thing is to import PyTorch and other necessary modules.

```
# Import necessary packages
import torch
from torch import Tensor
import torch.nn as nn
import torch.nn.functional as F
import torch.optim as optim
import random
import numpy as np
import json
import math
```

In this part, the structure of the feedforward neural network is defined. Refer to the PyTorch document for more details.

```
# Define model structures and functions
class Net(nn.Module):
    def __init__(self):
        super(Net, self).__init__()
        # Define a fully connected layers model with three inputs
(frequency, flux density, duty ratio) and one output (power loss).
        self.layers = nn.Sequential(
            nn.Linear(3, 15),
            nn.ReLU(),
            nn.Linear(15, 15),
            nn.ReLU(),
            nn.Linear(15, 15),
            nn.ReLU(),
            nn.Linear(15, 1),
        )

    def forward(self, x):
        return self.layers(x)


def count_parameters(model):
    return sum(p.numel() for p in model.parameters() if
p.requires_grad)
```

In this part, the dataset is loaded and pre-processed for the network training and testing. A small dataset containing triangular waveforms measured with N87 ferrite material under different frequency, flux density, and duty ratio values is used and can be downloaded from the MagNet GitHub repository under the "tutorial" section.

The provided code loads a dataset from a JSON file (`Dataset_tri.json`) and processes it for machine learning tasks. The dataset includes four variables: frequency (`Freq`), flux density (`Flux`), duty ratio (`Duty`), and power loss (`Power`). First, it applies logarithmic transformations to the `Freq`, `Flux`, and `Power` variables. The data is then reshaped into column vectors for each feature. These features are concatenated into a single array (`temp`), which is then converted into PyTorch tensors for the input data (`in_tensors`). The Power values are converted into output tensors (`out_tensors`). Finally, the code returns a PyTorch `TensorDataset`, containing both input and output tensors, ready for use in training models. The dataset is also prepared for potential future use, with commented-out code that saves the tensors as `.npy` files.

```
# Load the dataset

def get_dataset():
    # Load .json Files
    with open('/content/Dataset_tri.json','r') as load_f:
        DATA = json.load(load_f)

    Freq = DATA['Frequency']
    Flux = DATA['Flux_Density']
    Duty = DATA['Duty_Ratio']
    Power = DATA['Power_Loss']

    # Compute labels
    # There's approximalely an exponential relationship between
Loss-Freq and Loss-Flux.
    # Using logarithm may help to improve the training.
    Freq = np.log10(Freq)
    Flux = np.log10(Flux)
    Duty = np.array(Duty)
    Power = np.log10(Power)

    # Reshape data
    Freq = Freq.reshape((-1,1))
    Flux = Flux.reshape((-1,1))
    Duty = Duty.reshape((-1,1))

    print(np.shape(Freq))
    print(np.shape(Flux))
    print(np.shape(Duty))
    print(np.shape(Power))

    temp = np.concatenate((Freq,Flux,Duty),axis=1)

    in_tensors = torch.from_numpy(temp).view(-1, 3)
    out_tensors = torch.from_numpy(Power).view(-1, 1)

    # # Save dataset for future use
    # np.save("dataset.fc.in.npy", in_tensors.numpy())
    # np.save("dataset.fc.out.npy", out_tensors.numpy())

      return              torch.utils.data.TensorDataset(in_tensors,
out_tensors)
```

In this part, the training and testing procedure of the network model are programed. The loaded dataset is randomly split into the training set, validation set, and test set. The output of the training is the state dictionary file (`.sd`) containing all the trained parameter values.

The training process in the code starts by setting a fixed random seed for reproducibility and defining key hyperparameters such as the number of epochs, batch size, learning rate, and learning rate decay. The model is trained using the Adam optimizer and MSE as the loss function. The dataset is split into training, validation, and test sets (60, 20, and 20%, respectively). During each epoch, the model trains on batches from the training dataset, adjusting its weights using backpropagation based on the computed loss. The learning rate decays every 100 epochs by a factor of 0.5. After each epoch, the performance of the model on the validation set is computed. Every 200 epochs, the training and validation losses are printed for monitoring progress. Once the training is completed, the parameters of the model are saved.

In the evaluation phase, the model is switched to evaluation mode, and its performance on the test set is assessed. The predicted outputs and the true labels are collected and the MSE loss is computed on the test set. The results are then transformed back from the logarithmic scale to the original scale by applying the inverse logarithm. The relative error, average relative error, root mean square (RMS) error, and maximum error are calculated to measure the accuracy of the model. These metrics provide a detailed view of how well the model generalizes to unseen data, with the relative error giving an overall measure of prediction accuracy and the RMS and maximum errors highlighting the magnitude of deviations from the true values.

```
# Config the model training

def main():

    # Reproducibility
    random.seed(1)
    np.random.seed(1)
    torch.manual_seed(1)
    torch.backends.cudnn.deterministic = True
    torch.backends.cudnn.benchmark = False

    # Hyperparameters
    NUM_EPOCH = 1000
    BATCH_SIZE = 128
    DECAY_EPOCH = 100
    DECAY_RATIO = 0.5
    LR_INI = 0.02

    # Select GPU as default device
    device = torch.device("cuda")

    # Load dataset
    dataset = get_dataset()

    # Split the dataset
    train_size = int(0.6 * len(dataset))
    valid_size = int(0.2 * len(dataset))
    test_size = len(dataset) - train_size - valid_size
    train_dataset, valid_dataset, test_dataset =
torch.utils.data.random_split(dataset, [train_size, valid_size,
test_size])
    kwargs = {'num_workers': 0, 'pin_memory': True,
'pin_memory_device': "cuda"}
    train_loader = torch.utils.data.DataLoader(train_dataset,
batch_size=BATCH_SIZE, shuffle=True, **kwargs)
```

```
    valid_loader = torch.utils.data.DataLoader(valid_dataset,
batch_size=BATCH_SIZE, shuffle=False, **kwargs)
    test_loader = torch.utils.data.DataLoader(test_dataset,
batch_size=BATCH_SIZE, shuffle=False, **kwargs)

    # Setup network
    net = Net().double().to(device)

    # Log the number of parameters
    print("Number of parameters: ", count_parameters(net))

    # Setup optimizer
    criterion = nn.MSELoss()
    optimizer = optim.Adam(net.parameters(), lr=LR_INI)

    # Train the network
    for epoch_i in range(NUM_EPOCH):

        # Train for one epoch
        epoch_train_loss = 0
        net.train()
        optimizer.param_groups[0]['lr'] = LR_INI* (DECAY_RATIO **
(0+ epoch_i // DECAY_EPOCH))

        for inputs, labels in train_loader:
            optimizer.zero_grad()
            outputs = net(inputs.to(device))
            loss = criterion(outputs, labels.to(device))
            loss.backward()
            optimizer.step()

            epoch_train_loss += loss.item()

        # Compute Validation Loss
        with torch.no_grad():
            epoch_valid_loss = 0
            for inputs, labels in valid_loader:
                outputs = net(inputs.to(device))
                loss = criterion(outputs, labels.to(device))

                epoch_valid_loss += loss.item()

        if (epoch_i+1)%200 == 0:
          print(f"Epoch {epoch_i+1:2d} "
              f"Train {epoch_train_loss / len(train_dataset) *
1e5:.5f} "
              f"Valid {epoch_valid_loss / len(valid_dataset) *
1e5:.5f}")
```

```
    # Save the model parameters
    torch.save(net.state_dict(), "/content/Model_FNN.sd")
    print("Training finished! Model is saved!")


    # Evaluation
    net.eval()
    y_meas = []
    y_pred = []
    with torch.no_grad():
        for inputs, labels in test_loader:
            y_pred.append(net(inputs.to(device)))
            y_meas.append(labels.to(device))

    y_meas = torch.cat(y_meas, dim=0)
    y_pred = torch.cat(y_pred, dim=0)
    print(f"Test Loss: {F.mse_loss(y_meas, y_pred).item() /
len(test_dataset) * 1e5:.5f}")

    yy_pred = 10**(y_pred.cpu().numpy())
    yy_meas = 10**(y_meas.cpu().numpy())

    # Relative Error
    Error_re = abs(yy_pred-yy_meas)/abs(yy_meas)*100
    Error_re_avg = np.mean(Error_re)
    Error_re_rms = np.sqrt(np.mean(Error_re ** 2))
    Error_re_max = np.max(Error_re)
    print(f"Relative Error: {Error_re_avg:.8f}")
    print(f"RMS Error: {Error_re_rms:.8f}")
    print(f"MAX Error: {Error_re_max:.8f}")

if __name__ == "__main__":

    main()
```

The results demonstrate a significant improvement in the model performance over time. Initially, at epoch 200, the training and validation losses are relatively high (around 1.1), indicating that the model is still in the early stages of learning. As training progresses, both losses decrease steadily, with the training loss reaching 0.23211 and the validation loss stabilizing around 0.23384 by the 1000th epoch. This suggests effective learning and generalization. After training, the test loss is low at 0.01655, indicating that the model generalizes well to unseen data. However, the relative error of 3.28%, RMS error of 4.12%, and maximum error of 15.13% suggests that, while the model performs well overall, there are some discrepancies between the predicted and actual values, particularly with a few larger errors. These discrepancies may be due to the complexity of the data or limitations in the model. More details and a tutorial can be found in [11].

```
(9754, 1)
(9754, 1)
(9754, 1)
(9754,)
Number of parameters:  556
Epoch 200 Train 1.11248 Valid 1.05782
Epoch 400 Train 0.36304 Valid 0.37167
Epoch 600 Train 0.26303 Valid 0.24994
Epoch 800 Train 0.23845 Valid 0.23868
Epoch 1000 Train 0.23211 Valid 0.23384
Training finished! Model is saved!
Test Loss: 0.01655
Relative Error: 3.28383641
RMS Error: 4.12429436
MAX Error: 15.12973716
```

2.7 Tools and Software

This section provides helpful resources related to machine learning (ML) in power electronics, including open access dataset, tools, and software for integrating ML in power electronics applications.

2.7.1 Datasets

The following are useful datasets and resources related to power electronics:

I. IEEE DataPort

Website: https://ieee-dataport.org/

A repository of datasets across multiple domains, including power electronics, motor drives, and energy systems.

II. Open Power System Data

Website: https://open-power-system-data.org/

Provides datasets for power system analysis and modeling, including electricity generation and consumption.

III. Kaggle

Website: https://www.kaggle.com/datasets

A platform offering numerous datasets, including those related to energy and power electronics.

2.7.2 Programming Languages

I. Python—scikit-learn

Purpose: Traditional machine learning.

Key Features: Provides implementations of algorithms for classification, regression, clustering, dimensionality reduction, and data preprocessing.

II. Python—TensorFlow and Keras

Purpose: Deep learning.

Key Features: TensorFlow is an comprehensive library for building neural networks and ML applications. Keras, its high-level API, enables rapid development and prototyping.

III. Python—PyTorch

Purpose: Deep learning.

Key Features: A flexible deep learning framework, favored in research for its dynamic computation graphs and intuitive design.

IV. MATLAB—Statistics and Machine Learning Toolbox: Provides algorithms and functions for statistical analysis, classification, regression, clustering, and more. Key features include support for various models (e.g., decision trees, SVMs, neural networks).

V. MATLAB—Deep Learning Toolbox: Offers tools for designing, training, and implementing deep neural networks, including pre-built layers and functions for common deep learning tasks.

VI. MATLAB—Computer Vision Toolbox: Useful for image processing and computer vision applications, including object detection, classification, and feature extraction using machine learning techniques.

VII. MATLAB—Reinforcement Learning Toolbox: Provides algorithms for training agents in reinforcement learning environments, enabling the design and training of intelligent systems.

VIII. MATLAB—Automated Machine Learning (AutoML): Streamlines the process of selecting features, choosing algorithms, and tuning hyperparameters, making it easier to build machine learning models.

IX. MATLAB—Parallel Computing Toolbox: Speeds up machine learning tasks by utilizing multiple cores or GPUs for training large models.

2.7.3 Tools for Model Tuning and Hyperparameter Optimization

Several tools and frameworks are available for tuning models and optimizing hyperparameter in machine learning. Each provides distinct features suited to different use cases.

I. Ray Tune: A scalable tool that supports distributed hyperparameter search, making it ideal for large-scale applications.
II. Optuna: An efficient, automatic hyperparameter optimization framework with pruning capabilities to accelerate the search process.
III. Google Vizier: A black-box optimization service used for tuning machine learning models at scale.
IV. Microsoft NNI (Neural Network Intelligence): Offers automated hyperparameter tuning and neural architecture search.

References

1. Xia H, Zhang Y, Chen M, Luo D, Lai W, Wang H (2024) Capacitor parameter estimation based on wavelet transform and convolution neural network. IEEE Trans Power Electron 39(11):14888–14897
2. Son J, Kim C, Jeong M (2022) Unsupervised learning for anomaly detection of electric motors. Int J Precis Eng Manuf 23:421–427
3. Jakobeit D, Schenke M, Wallscheid O (2023) Meta-reinforcement-learning-based current control of permanent magnet synchronous motor drives for a wide range of power classes. IEEE Trans Power Electron 38(7):8062–8074
4. Gamal Al-kaf HA, Hakami SS, Halabi LM, Lee K-B (2022) Model predictive current control using single layer neural network for PMSM drives. In: IEEE energy conversion congress and exposition (ECCE). Detroit, MI, USA
5. Kim S-H, Yoo D-Y, An S-W, Park Y-S, Lee J-W, Lee K-B (2020) Fault detection method using a convolution neural network for hybrid active neutral-point clamped inverters. IEEE Access 8:140632–140642
6. Álvarez Antón JC, García Nieto PJ, Blanco Viejo C, Vilán Vilán JA (2013) Support vector machines used to estimate the battery state of charge. IEEE Trans Power Electron 28(12):5919–5926
7. Yang Z, Zeng J, De La Cruz A (2024) Learning and outperforming the model predictive control with a linear ε-support vector machine for power converter. IEEE Trans Ind Electron 71(1):5–13
8. Gasper P, Schiek A, Smith K, Shimonishi Y, Yoshida S (2022) Predicting battery capacity from impedance at varying temperature and state of charge using machine learning. Cell Rep Phys Sci 3(12):101184
9. Elsamanty M, Ibrahim A, Salman WS (2023) Principal component analysis approach for detecting faults in rotary machines based on vibrational and electrical fused data. Mech Syst Signal Process 200:110559
10. Schenke M, Wallscheid O (2021) A deep Q-learning direct torque controller for permanent magnet synchronous motors. IEEE Open J Ind Electron Soc 2:388–400
11. Chen M et al (2024) MagNet challenge for data-driven power magnetics modeling. IEEE Open J Power Electron

Chapter 3
Integrating Machine Learning into Real-Time Power Electronics Applications on DSPs

Abstract This chapter provides an in-depth analysis of the key considerations and challenges involved in integrating ML into real-time power electronics and motor drive applications. It begins by addressing challenges such as limited data availability and the need for rapid tuning and control speeds, which are crucial in high-performance power electronic systems. These factors often complicate the deployment of ML models, necessitating solutions that deliver accuracy, speed, and efficiency. Following this, this chapter presents practical steps for deploying ML in real-time power electronics environments, with a focus on digital signal processors (DSPs). DSPs are widely implemented in real-time applications due to their optimized computational capabilities, making them particularly well-suited for handling the rapid processing and control demands of power electronics and motor drive systems. This chapter then presents strategies for integrating ML into DSPs, including both direct ML implementations and hybrid approaches that combine traditional control algorithms with ML components. Various effective neural network algorithms that have been successful applied in DSP applications are examined, along with insights into their specific implementation techniques. Finally, this chapter provides comprehensive guidelines and considerations for implementing ML on DSPs, covering stages from data collection to model deployment.

3.1 Challenges of Integrating ML in Real-Time Power Electronics Applications

The implementation of ML in power electronics presents unique challenges and characteristics that are distinct from other domains such as image classification. Power electronic systems require extremely high tuning speeds for control and exhibit high sensitivity in condition monitoring. Additionally, ML-based controllers often struggle with the inherent complexity of their models and the scarcity of suitable training datasets. Since their performance depends strongly on large, high-quality data, any imbalance or deficiency in the training samples can noticeably degrade control accuracy and reliability. In addition, deep neural networks (DNNs) and other

H. A. G. Al-kaf and K.-B. Lee, *Machine Learning Integration in Power Electronics*,
https://doi.org/10.1007/978-981-95-3845-4_3

types of reinforcement learning (RL) can strain system resources, necessitating high-speed processors for real-time applications. The fine-tuning of training parameters for optimal performance is often a complex and time-consuming process. Therefore, these specific demands require tailored ML approaches to meet the rigorous performance requirements of power electronics.

In electric machine drives, the need for real-time inference of ML algorithms has become increasingly challenging. For example, modern motor control applications operate at frequencies ranging from 10 to 40 kHz, necessitating that the available computation time for each control loop be as brief as 25–100 μs. This tight timeframe, after accounting for ADC sampling, signal scaling, and software-based protection logic, leaves minimal time for ML inference [1].

Fault detection in power electronics and motor drives is a complex and critical task, often complicated by the presence of various faults that can occur under diverse operating conditions. Collecting data under these conditions can be dangerous and may damage the system. Many ML models, especially CNNs, have significant computational demands, involving numerous layers and operations that require extensive processing time. This high computational cost becomes particularly problematic in real-time applications, where fast response times are crucial to prevent system damage or failure. Additionally, the training and inference phases of such models can be resource-intensive, demanding considerable memory and computational power that may not be available in the embedded systems commonly used in power electronics and motor drive environments.

Deploying ML methods in C++ for DSPs or in VHDL for field-programmable gate arrays (FPGAs) presents substantial challenges due to the fundamental differences between high-level ML frameworks such as Python-based TensorFlow and MATLAB, and low-level hardware-oriented programming. While Python and MATLAB provide extensive ML libraries with highly optimized numerical operations, these frameworks remain incompatible with resource-constrained embedded systems. In C++, implementing ML models requires careful optimization to address memory constraints, computational efficiency, and real-time processing requirements. This process often involves the use of fixed-point arithmetic and hardware-friendly approximations to ensure feasibility within embedded environments. Similarly, deploying ML algorithms on FPGAs using VHDL requires a hardware-aware design approach, which involves managing parallelism effectively, optimizing resource allocation, and low-latency inference execution. Furthermore, the lack of direct compatibility with widely used ML frameworks, the limited computational resources of embedded platforms, and the extensive debugging efforts required for deployment increase the complexity of integrating ML into such systems. Consequently, adapting ML models for embedded hardware remains an intricate and labor-intensive process, demanding expertise in both ML algorithms and hardware design methodologies.

3.2 Practical Steps for Deploying ML in Real-Time Power Electronics Applications on DSPs

In recent years, significant advancements have been made in implementing ML in power electronics, particularly through its deployment on DSPs and FPGAs. DSPs have been widely used for controlling electric machine drives by leveraging their optimized processing capabilities for real-time signal processing tasks. However, many existing DSP-based systems employ relatively shallow neural network structures and operate with slower pulse-width modulation (PWM) cycles [2]. As demand grows for more sophisticated ML applications, especially those used in real-time environments, DSPs face challenges in achieving the required inference speeds. This is particularly true for deep learning models, which must execute within microseconds to match the control frequencies typical of modern motor control systems.

Meanwhile, FPGAs have emerged as a highly promising alternative for implementing ML algorithms in electric drives. Their ability to provide low latency, high throughput, and flexibility allows for the execution of deeper neural networks with real-time performance. A key advantage of FPGAs is their parallel processing capability, enabling them to handle multiple operations simultaneously. This significantly boosts performance for tasks like deep learning inference and real-time control applications. FPGAs can be customized to optimize the inference of ML models while simultaneously managing various sensor data inputs, such as current and voltage measurements. Furthermore, the cost-effectiveness and low power consumption of FPGAs make them suitable for a broad range of applications, from consumer appliances to electric vehicles. With advanced architectures, such as system-on-chip (SoC) designs that combine processing units and programmable logic, FPGAs are well positioned to enhance the performance and efficiency of smart control systems in electric drives, thereby accelerating innovation in the field.

However, various recent studies have successfully implemented ML algorithms on DSP platforms, demonstrating their effectiveness in practical applications. The steps for implementing ML in power electronics on a real-time DSP are shown in Fig. 3.1. The implementation of ML in power electronics and motor drives in DSPs can be online or offline, depending on the purpose of the implementation. For example, ML can be applied offline to optimize the weights of a MPC cost function without the need for direct implementation on a DSP. This optimization can be achieved using a lookup table or optimized weights and biases. In lookup table approaches, once the ML algorithm completes training, the optimal values are stored for future use. For example, a lookup table can be used to fine-tune the parameters of the adaptive frequency response and PI controller. In cases involving ANN structures, the optimized weights and biases, together with a simplified mathematical model, can be implemented in a DSP without the computational overhead of training the algorithm. For instance, after conducting offline training to solve a specific problem, the optimized parameters are applied to the DSP. Nevertheless, multiple essential considerations must be addressed throughout this implementation phase. Initially, it is crucial to verify that the optimized weights and biases derived from Python or MATLAB

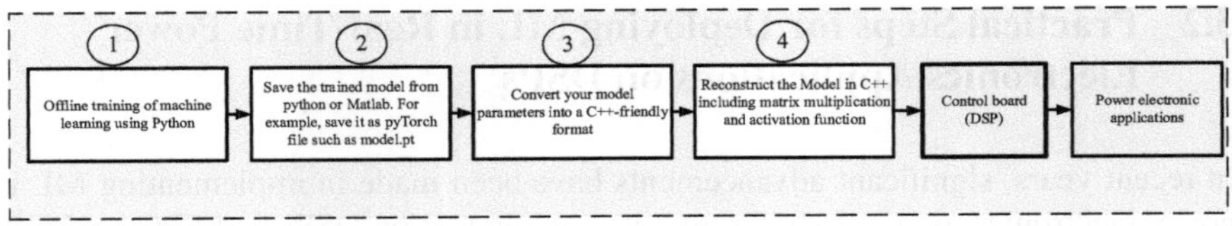

Fig. 3.1 Block diagram of the machine learning workflow in a DSP

simulations are compatible with the target system, such as C programming for DSP platforms. The activation function, hidden layers, and neuron structure employed in computational models must be precisely implemented in the experimental setup. Ultimately, performance requires comprehensive assessment across both simulation and experimental environments to maintain consistency. ANN structures can be streamlined by eliminating insignificant neurons or implementing low-computation activation functions.

3.3 Strategies for Integrating Machine Learning

In this section, two methods are presented to implement ML in real-time power electronics applications including direct ML and hybrid ML methods.

3.3.1 Direct Machine Learning (DML) Integration

The DML approach in power electronics and motor drive applications relies on ML algorithms to process input data and to make decisions or control actions. In these applications, ML algorithms are typically trained on large datasets that represent various operating conditions of the system. Once trained, the algorithm can predict or optimize outcomes such as motor speed, torque, fault detection, and energy efficiency in real time.

For example, in [3], an ANN was implemented in the control system of an induction motor. Induction motors are nonlinear, time-varying systems that are challenging to control due to difficulties in measuring state variables and variations such as rotor resistance changes from temperature fluctuations and inductance saturation. By using an ANN, the motor speed control system can achieve high performance by efficiently estimating state variables and compensating for system nonlinearities. The inherent ability the ANN to manage time-varying nonlinearities makes it a suitable controller. To train the ANN for the induction motor drive system, a conventional FOC system is first implemented to gather training data, including several forward and reverse speed step commands. Although increasing the number of neurons in the network can reduce the output error, beyond a certain limit it can cause overfitting, where the network performs well on training data but poorly on test data. To address this, the ANN is re-trained using input data derived from the experimental setup. If the results fall outside the expected range, the desired output for each set of recorded experimental input data is recalculated using FOC. The new training data consists of experimental data combined with data from the initial training stage. This process is repeated until satisfactory speed and torque results are achieved, indicating that the maximum absolute difference between the FOC system and the ANN system is less than 8%. The ANN controller is then implemented on a DSP development module with a Texas Instruments TMS320C30 CPU. Although the TMS320C30 lacks a tangent sigmoid instruction, the function is still implemented using C language math libraries, which can be computationally expensive. A more efficient approach uses a lookup table to calculate the tangent sigmoid function, reducing processor time while minimizing error.

In [4], an ANN was designed to dynamically calibrate the parameters of an adaptive frequency response (AFPR) controller for a stepper motor to compensate for discrepancies between the theoretical model and the actual motor especially when motor characteristics change. The ANN calibrating the AFPR controller uses four inputs: the reference current (i_{ref}), the control output (u), the current tracking error (e), and the change in error ($^{\Delta}e$). These inputs help the ANN dynamically adjust the AFPR parameters including the proportional gain ($P_{PR\mathrm{i}}$), the resonant gain (K_{Ri}), and the quality factor (Q_{Ri}) to minimize the tracking error. Using an online back-propagation algorithm with gradient descent, the ANN continuously tunes these parameters, which are then stored in an offline lookup table to ensure consistent and optimal motor performance across varying conditions. The calibration process was performed before the official motor operation to select the optimal parameters for the AFPR controller, minimizing the current error. The ANN training time is set to 150 s at each speed in the frequency region from 1 to 300 Hz. A block diagram of the proposed controller is shown in Fig. 3.2. A DSP TMS320F2812 was used to apply the proposed algorithms. The sampling time was implemented at 25 μs.

In [5], a novel approach was proposed for controlling inverter-based distributed generation (DG) units by using an ANN for real-time system identification and control. The key idea revolves around using an ANN to address the challenges in estimating interfacing parameters and grid voltage without the need for direct grid-voltage sensors. A two-layer neural network interfacing parameter identifier (NNIPI) was used to estimate the unknown interfacing parameters of the inverter by minimizing the parameter estimation error using an iterative gradient projection algorithm. This approach ensures that the interfacing parameters are accurately identified in real-time. Another two-layer neural network grid-voltage estimator (NNGVE) is used to estimate the grid voltage at the point of common coupling (PCC). The NNGVE uses a steepest descent adaptation algorithm to dynamically update its estimates, making it robust to changes in the grid voltage. In addition, a discrete Lyapunov stability analysis was presented to ensure that the ANN estimations are reliable and that the system remains stable. This analysis helps properly tune the ANN to guarantee convergence of the system control algorithms. A TMS320C30 DSP was chosen as an embedded platform. The execution time of the current control

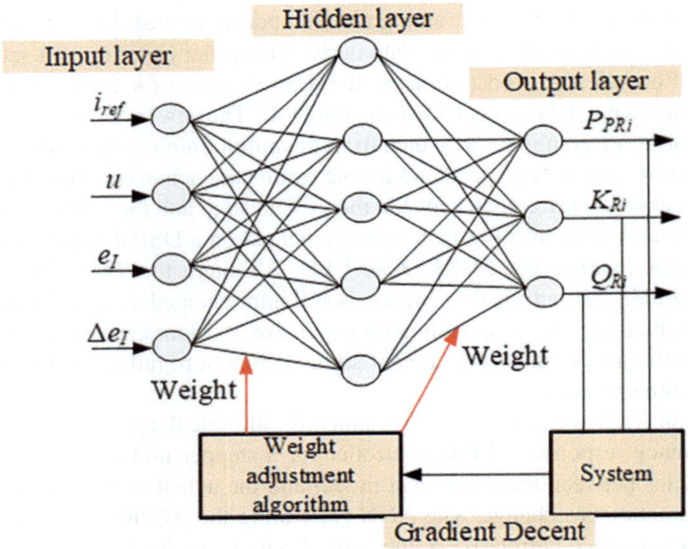

Fig. 3.2 NN structure of DML for dynamic calibration of the parameters of the AFPR [4]

interrupt routine is about 130 s. Subsequently, a control period of 150 s is selected. With these setting, a safe CPU load coefficient of 86% and a switching frequency of 6.7 kHz have been obtained. The ANN control scheme imposes only a modest computational load on a DSP, enabling operation at high switching frequencies.

In [6], a DC bus capacitance identification method based on the ANN algorithm was proposed. The proposed method requires no additional current sensors on the DC bus side of the rectifier. Moreover, it can be implemented using the existing resources of most digital controllers in power electronic systems. The ANN approach is employed in EV charging systems to estimate variations in the DC-bus capacitance. The ANN employs the phase-a voltage (e_a) and current (i_a), and the dc bus ripple (U_{dc}). Since the experimental data collected in the hardware circuit using the DSP TMS320F28335 contain random noises and outliers, preprocessing and filtering are required. They are preprocessed with recursion average filtering to the outliers, and the five points-thrice smoothing technique is employed to remove random noises as shown in Fig. 3.3. This step transforms the experimental data to closely match the training simulation data, a crucial process for online DC-bus capacitance identification. The preprocessed data are then used for capacitance estimation. To enhance identification reliability, the capacitance value is supplied to the controller only when its variation stabilizes within 3%.

To sum up, several strategies have been proposed to successfully implement robust DML algorithms in DSP systems. After training with simulation data, it is recommended to use experimental datasets for further model validation. Filtering and smoothing techniques are essential for removing noise and outliers from the dataset.

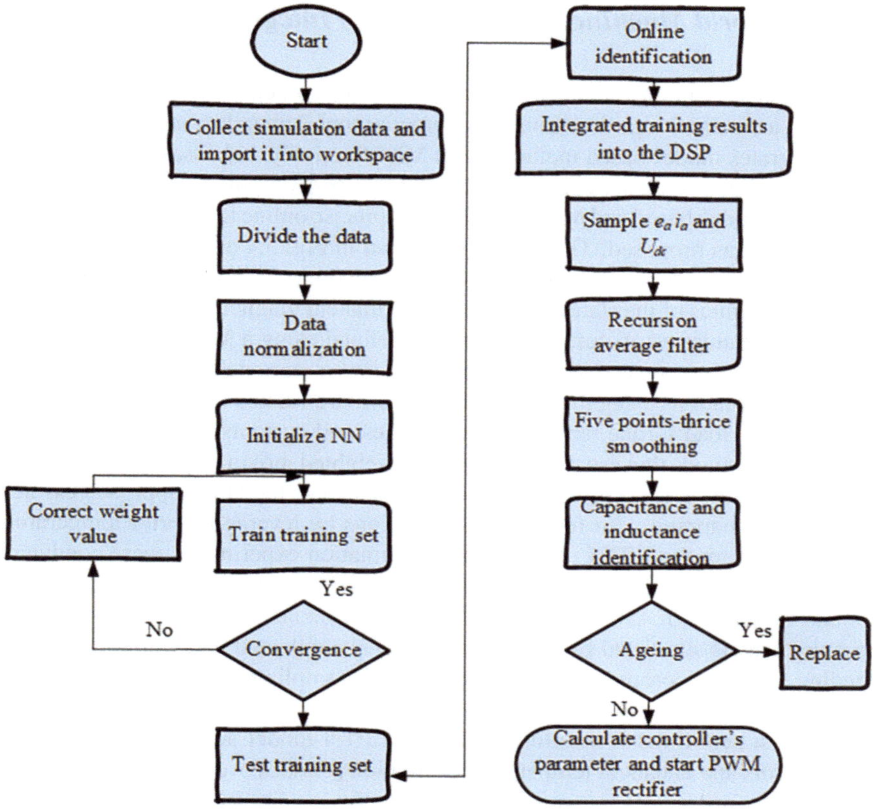

Fig. 3.3 Flowchart of DML for dc-bus capacitance identification [6]

Additionally, low-computation training algorithms such as the iterative gradient projection algorithm and the steepest descent adaptation algorithm can effectively ensure that the model remains reliable and robust when applied to a DSP. Furthermore, using a small number of neurons and activation functions such as ReLU and hyperbolic tangent, or utilizing a lookup table for tangent sigmoid function values can help reduce the complexity of ANNs. Using constant weights in some layers and offline-optimized weights reduces the computation time of the ML algorithm without compromising performance. It is important to ensure high-accuracy performance and conduct a stability analysis, such as Lyapunov stability analysis. Stability analysis ensures that the system remains stable and reliable during operation. Implementing ML to generate a lookup table can also significantly reduce computation time. Furthermore, various strategies for DML are discussed in the following sections.

3.3.2 Hybrid Machine Learning (HML) Integration

DML methods can have challenges such as the time-intensive data collection and training, as well as significant online computational demands. An HML approach that integrates model-based methods with ML effectively addresses the complexity of generated data, enhances performance, and reduces computational time. In [7], an integrated model-based and ML method for the precise online temperature estimation of PMSMs was proposed. The process is shown in Fig. 3.4 and begins with developing a simplified lumped parameter thermal network (LPTN) model to capture the thermal dynamics of the stator winding and permanent magnet. This model provides initial temperature estimates, which are then refined using a ML approach. Specifically, an ANN compensator is designed to account for unmodeled characteristics and parameter variations using inputs such as temperature-related information, loss data, motor speed, load torque, and d-q axis voltages and currents. Unlike existing ML estimation methods that use the exponentially weighted moving average technique to gather information by expanding the input dimensions, the HML approach extracts an effective mapping with fewer data dimensions by leveraging prior temperature knowledge from the LPTN model. Online estimation experiments were conducted on a standard industrial microcontroller (DSP TMS320F28335), which operates at a main frequency of 150 MHz. In real-time experiments, the acquisition of online sampled data is simulated by importing previously collected data into the DSP and feeding it into the proposed method during each sampling period. This effectively replicates the use of real-time sampled data.

In [8], a neural network informed by the LPTN model addresses unaccounted model errors and enhances temperature estimation at nodes overlooked by the highly abstract low order LPTN, even when training data is limited. This hybrid model provides two outputs: core PMSM temperature estimation based on LPTN nodes, and extended temperature estimation for additional locations such as bearings and end windings. The LSTM focuses on residual model uncertainties and expands temperature estimation without needing to model all of the thermal dynamics. This approach simplifies the framework, reduces data requirements, and leverages adaptive learning of the LSTM to enhance temperature estimation across various motor components. Unlike DML methods that require training datasets exceeding 100 h, the HML combines the simplicity of LPTN for initial modeling with the advanced capabilities of LSTM for refinement. It achieves a low estimation error using a compact dataset of just 23.8 h. The efficiency of this model showcases its robustness and facilitates its practical adoption.

Another application of HML is shown in Fig. 3.5, which presents a block diagram of the proposed HML for controlling power flow in a multi-active bridge (MAB) converter [9]. An ANN model is introduced to replace the traditional numerical solver, acting as the feedforward component in a feedback control loop. This model takes the desired power flow as its input and predicts the phase angles necessary to achieve the target power flow. The principle of the transfer learning method is used for the MAB to control power flow.

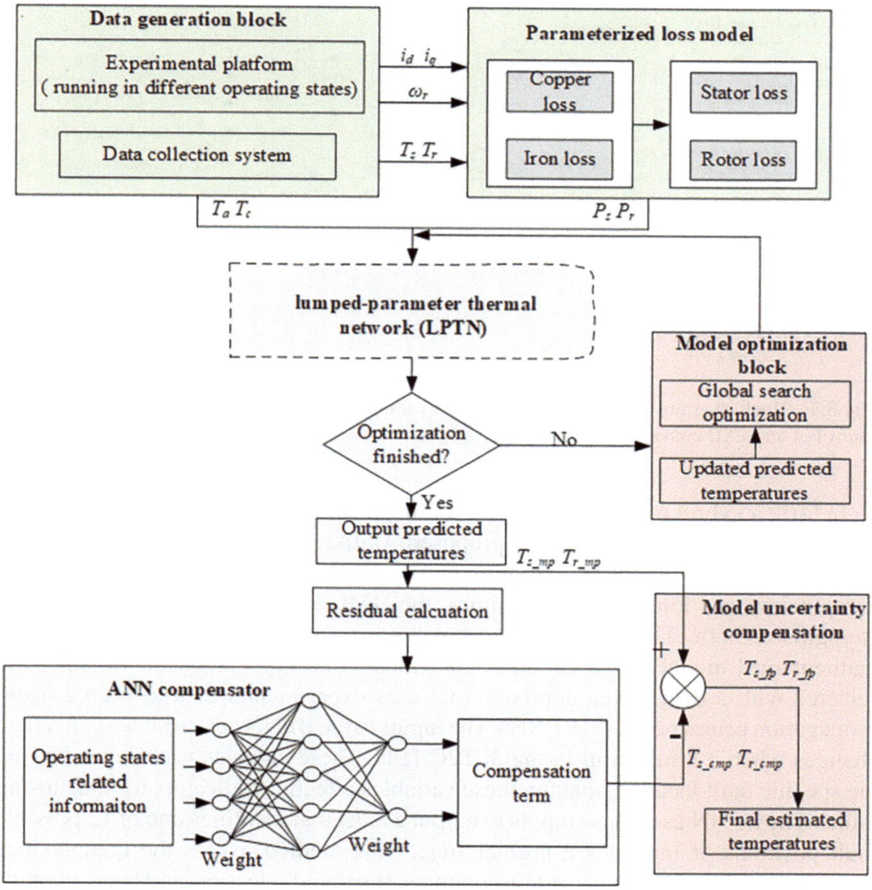

Fig. 3.4 Block diagram of a hybrid model for temperature estimation of a PMSM [7]

In this approach, the ANN is pre-trained using data derived from theoretical calculations in the trapezoidal mode. Then it is fine-tuned with data obtained from experimental measurements. This allows the model to account for nonidealities and component variations. Since models built from ideal and experimental data share common features, the knowledge from traditional modeling can be reused and transferred between different scenarios. Then the accuracy of the ANN obtained through transfer learning is compared to the accuracy of an ANN trained using the traditional method with a large dataset. The rapid advancements of ML frameworks like PyTorch and TensorFlow, along with numerous open-source examples available online have significantly accelerated the development of models. For power flow modeling, an FNN is trained using TensorFlow and then converted to TensorFlow Lite, enabling seamless integration between the software and hardware systems. The model is then encoded and deployed onto an ANN accelerator for phase prediction.

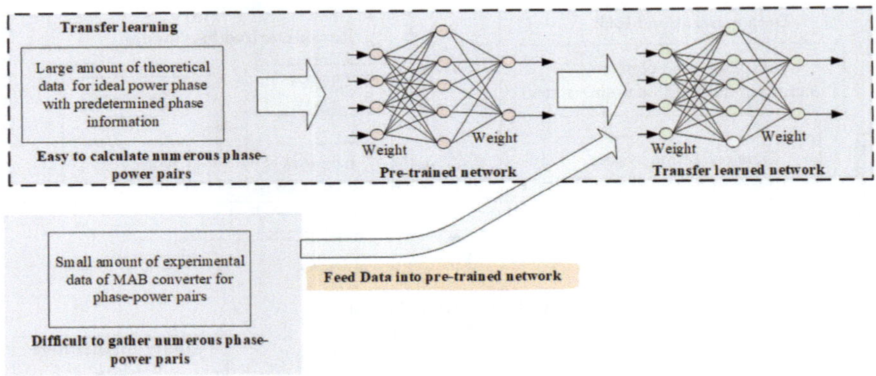

Fig. 3.5 Block diagram of the hybrid model and transfer learning for feedforward power flow control of an MAB converter [9]

In [10], a hybrid model-ML method for diagnosing open-circuit faults in a t-type three-level converter (3LT2C) was proposed. Unlike the model-driven method that requires complex rules to classify faults, and the DML method that demands intensive computation and longer diagnosis times, the HML approach aims to combine the strengths of both. The HML method combines a model-based approach where a mathematical model of the circuit is constructed and key diagnostic variables are selected with a data-driven approach that uses experimental data to train a back-propagation neural network (BPNN). The inputs to the BPNN are variables that reflect changes when a bridge arm in the 3LT2C fails. These changes vary depending on the specific fault location, making these variables effective indicators for diagnosing faults. The BPNN uses these inputs to output a label that identifies one of 12 possible fault positions or indicates a normal state. This method avoids the complexities associated with fault analysis, rule assignment, threshold selection, and large amounts of data to support calculations that impact the training speed. Thus, the diagnosis of single power device faults in traditional three-phase inverters can be achieved within several switching cycles and successfully implemented in real time using a DSP TMS320F28379D. The identification process can be completed in one sampling cycle of experimental data, taking less than 2 ms.

Similarly, an HML approach was proposed to diagnose open-switch faults in power converters [11]. The proposed method involves two main phases: offline model analysis and online fault diagnosis. First, analytical models of power converters are developed to monitor device voltages and detect faults. Then potential fault diagnosis variables are selected from these models and training data is collected. These variables are optimized through trial and error with training data to minimize the ANN computation, and an ANN is trained to map these variables to fault conditions. For online diagnosis, fault diagnosis variables are calculated using analytical models, and the trained ANN identifies faults in real-time. The model-driven approach speeds up diagnosis, reduces data-driven inputs, and simplifies learning, while the ML component automates rule-making and threshold selection, which is crucial for complex circuits.

3.4 Effective Neural Network Methods for DSPs

ANNs are widely implemented in DSPs due to their ability to efficiently handle real-time processing, nonlinear system modeling, and adaptive learning, making them superior to traditional ML methods. Unlike other approaches that rely on iterative optimization and heavy computations, an ANN can be pre-trained offline. This allows optimal weights and biases to be extracted and deployed for real-time inference with minimal computational overhead. Some DSP architectures are optimized for parallel processing and high-speed mathematical operations, making them well suited to ANNs, which depend on fast matrix multiplications and activation functions. Additionally, ANNs offer superior adaptability, allowing for online real-time parameter tuning using low computation training algorithms such as gradient descent or least mean squares (LMS) algorithms, enabling effective control in dynamic and uncertain environments such as motor drives, and power electronics. This section presents various ANN algorithms that have been effectively implemented in DSPs for power electronics applications.

3.4.1 Structured Multi-layer Network

The structured multi-layer network demonstrates strong effectiveness and has already been deployed on DSP platforms across a range of practical applications. It combines several layers, where each layer has its functions as shown in Fig. 3.6. In [12], the recurrent Hermite fuzzy neural network (RHFNN) was used to optimize the current angle command for a synchronous reluctance motor (SynRM) during transient states to enhance dynamic response. The RHFNN adjusts the current angle by computing a compensation term based on tracking errors and their rates of change. It operates through a structured six-layer network that includes an input layer, a membership layer, a rule and recurrent layer, a signed distance and Hermite layer, a consequent layer, and an output layer. The functions of each layer are described as follows.

The proposed architecture begins with an input layer that directly receives raw signals, such as stator current tracking errors and their derivatives. These inputs are then mapped into fuzzy domains in the membership layer, where Gaussian functions assign degrees of belonging to different fuzzy sets. Next, the rule–recurrent layer evaluates fuzzy inference rules while also incorporating feedback links that embed past information, thereby enhancing dynamic behavior. To simplify computations, the signed distance and Hermite layer reformulates the input space by calculating signed distances and expanding it with Hermite polynomials, enabling the

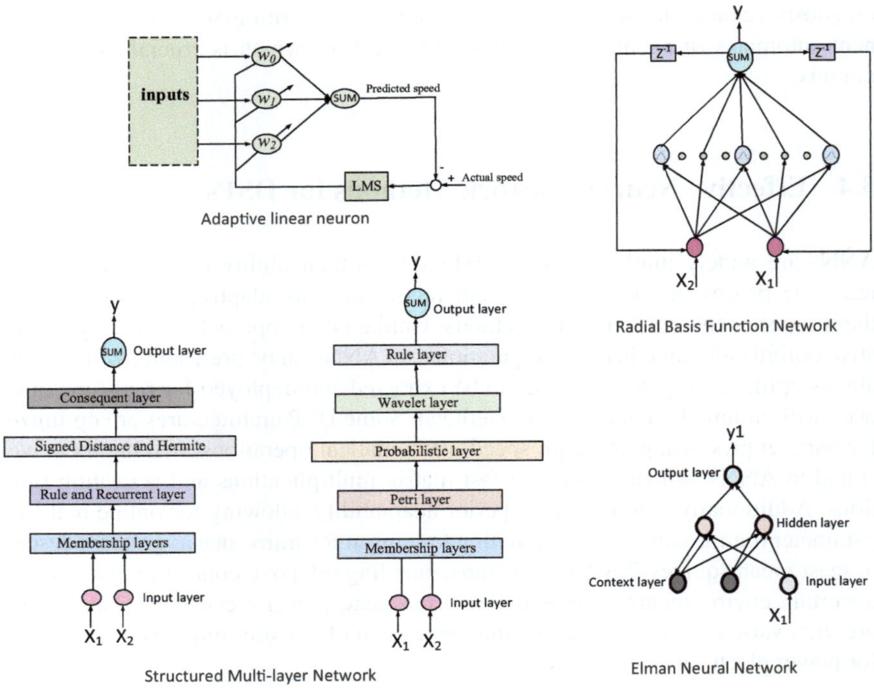

Fig. 3.6 Block diagram of neural network methods implemented on a DSP

system to represent nonlinear relationships more effectively. The consequent layer then integrates the transformed signals to generate corrective values for the current angle. Finally, the output layer consolidates these corrections to issue the real-time command for motor angle adjustment. Model parameters are continuously updated online through a gradient-descent learning mechanism, with the speed control loop executed at a 1 ms sampling period. Additionally, the execution time of the C program on the TMS320F28075 32-bit floating-point DSP (120 MHz) can be measured using the clock tool in the Texas Instruments Code Composer Studio v6.2 development environment. The total execution time of the proposed method is still less than 1 ms, which is the sampling interval of the speed control loop.

3.4.2 Elman Neural Network (ENN)

ENN exhibits distinct dynamical characteristics and has been widely applied in the identification and control of dynamical systems. Generally, the ENN can be considered to be a special kind of FNN with additional memory neurons and local feedback, as shown in Fig. 3.6. Owing to the context neurons and local recurrent connections between the context layer and the hidden layer, it has certain dynamical advantages over static NNs such as multi-layer perceptrons. However, the typical ENN struggles with approximating high-order dynamic systems and often has slow convergence, making it unsuitable for time-critical applications. In addition, the selection of activation functions significantly expands the search space of the ENN. To address these challenges, various studies have combined the learning capability of the ENN with wavelet decomposition, which is particularly effective for identifying and controlling dynamic systems. Wavelets, with their finite support and self-similarity, are well-suited for representing functions with local nonlinearities and variations. For instance, a recurrent wavelet Elman neural network (RWENN) can serve as an uncertainty observer, approximating lumped-parameter uncertainties and disturbances [13]. The stability of the closed-loop system was ensured through Lyapunov theory, with the TMS320F240 DSP employed for implementation.

3.4.3 Radial Basis Function Network (RBFN)

RBFN is a type of ANN used for tasks like classification and regression. It has three layers: an input layer, a hidden layer, and an output layer as shown in Fig. 3.6. The hidden layer is unique because it uses radial basis functions, often Gaussian, to process inputs based on their distance from a central point, or prototype. Each neuron in the hidden layer measures how close the input is to its prototype and outputs a value that decreases as the distance increases. The output layer combines these values to make predictions offering fast learning and good accuracy, especially for non-linear problems.

A recurrent RBFN (RRBFN) for the indirect FOC of an IM drive was proposed in [14]. The RRBFN acts as an uncertainty observer to estimate an unknown, nonlinear, time-varying function of lumped-parameter uncertainty online. It incorporates recurrent connections from the output back to the input, enabling dynamic feedback. An online learning algorithm based on backpropagation is used to recursively update both the weights and membership functions to minimize the error between the desired and actual rotor positions. This process adjusts the weights and Gaussian parameters according to specific update rules, allowing for real-time adaptive estimation and control. Additionally, adaptive learning algorithms for the RRBFN are derived using the Lyapunov stability theorem to train the network parameters online. All of the control algorithms are implemented in a TMS320C31 DSP-based control system.

3.4.4 Adaptive Linear Neuron (ADALINE)

ADALINE represents one of the earliest and simplest forms of ANNs, designed to handle linear data patterns. Its structure includes a set of inputs, associated weights, and an output computed as the weighted sum of those inputs. Learning is achieved through the LMS algorithm, which iteratively updates the weights to minimize the error between the predicted output and the desired target. As training progresses, the error gradually decreases, leading to more precise outputs. Because of this capability, ADALINE has found applications in areas such as pattern classification and signal analysis. In [15], the ADALINE system is used to compensate for errors in speed and position measurements caused by the encoder in an induction motor drive. It uses an ADALINE-based speed decomposer to extract components of the speed signal, such as DC, cosine, and sine variations, by setting these as inputs to the ADALINE network, as shown in Fig. 3.6. The ADALINE adjusts its weights to accurately estimate the magnitude of these components in real time. The goal is to correct errors in the encoder readings, particularly offset errors, which cause deviations in both speed and position. A TMS320F28335 DSP is used as the digital controller for executing the algorithm.

3.4.5 Deep Deterministic Policy Gradient (DDPG)

DDPG is a model-free, off-policy RL algorithm that combines the benefits of both policy-based and value-based methods. It utilizes an actor-critic structure where the actor updates the policy deterministically, while the critic estimates the value function. This approach is particularly effective for continuous action spaces, making it suitable for complex control tasks. In [16], the DDPG algorithm had an actor-critic structure, which is used to control a DC/DC buck converter, as shown in Fig. 3.7. The training is performed offline through iterative PC simulations to avoid potential damage to the hardware setup. The environment is simulated using a simplified buck converter model, where the inputs include the input/output voltages, inductor current, and current reference provided by an external PI controller. The outputs of the actor network, which is responsible for generating the control actions (specifically, the duty cycle for the buck converter), are then implemented on a TMS320F28379D DSP for real-time control. Only the actor network is used in the real-time application, while the critic network is utilized during training. To enable real-time implementation, the control law accelerator (CLA) of the DSP is employed to perform the computationally intensive matrix calculations of the ANN. This strategy reduces the computational load on the main processor of the DSP, ensuring stable and efficient operation of the converter. By using the CLA, these matrix calculations can be offloaded from the main processor. The CLA is designed to perform these operations more efficiently because it can execute them in parallel or in a more optimized sequence compared to the main processor. For instance, while the CLA processes matrix multiplication, the main processor can handle other control tasks such as monitoring sensor inputs or adjusting pulse-width modulation (PWM) signals. The integration of the Python-based stable-baselines3 RL library with the PLECS power electronics simulation tool is hindered by the lack of direct connectivity between the two. To bridge this gap, a communication environment was created using a dynamic linked library (DLL) in PLECS, with C language serving as an intermediary to exchange data between

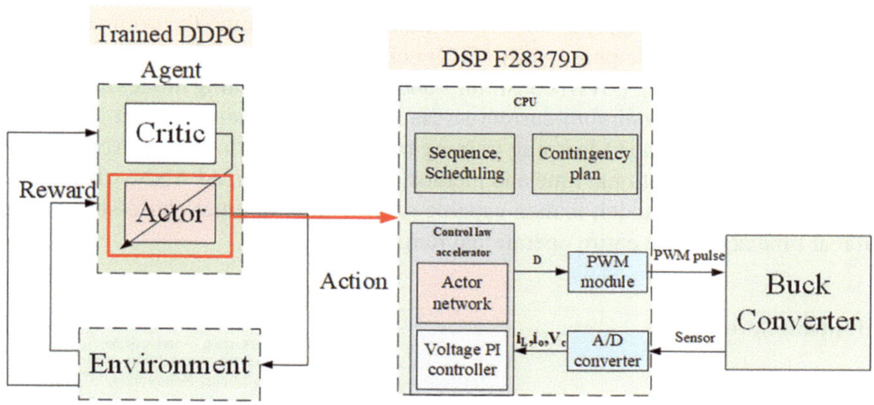

Fig. 3.7 Diagram of an RL-based (DDPG) controller implementation in a DC-DC buck converter

the RL agent and the simulation environment. Socket communication via localhost simplifies data exchange, allowing the actions of the agent to be sent to the environment and feedback states to be returned to the Python-based RL library. In real-time microcontroller implementation, the execution timing of the DLL block is crucial and must be adjusted to align with the PWM interrupt, considering the computational limitations of the microcontroller.

3.4.6 Q-Learning

Q-learning is a reinforcement learning algorithm extensively employed for sequential decision-making problems. It belongs to the class of model-free methods, as it does not require an explicit representation of the environment's dynamics. The algorithm constructs and iteratively updates a Q-function, stored in a Q-table, which estimates the expected cumulative reward associated with each state–action pair. Through continuous interaction with the environment and application of the Bellman optimality principle, the agent gradually refines its action-selection policy and converges toward an optimal strategy that maximizes long-term rewards. In [17], a Q-learning-based optimization scheme was proposed to reduce the power losses in a dual-active-bridge (DAB) converter. After the Q-learning algorithm is trained, a lookup table is typically generated to store the optimal action strategies. However, for converters with a wide operating range, this table can grow significantly in size, posing computational challenges due to the limited processing speed and memory capacity of the microcontroller. Furthermore, reducing the size of the Q-value table by increasing the intervals for the input voltage, output voltage, and transmitted power can lead to a loss in control accuracy. In [18], to reduce computation time, an RL + ANN-based minimum-current-stress scheme was implemented, as shown in Fig. 3.8. In the first stage, Q-learning is employed for offline training to determine the optimal control strategy based on triple-phase-shift (TPS) control. During the Q-learning process, zero-voltage-switching (ZVS) constraints and effective operating modes are considered, resulting in a minimum-current-stress scheme that maintains soft switching. In the second stage, the Q-learning training results are used to train an ANN, which helps reduce computational time and memory usage. The trained ANN, which functions as an implicit model, is then capable of providing optimal phase-shift angles in real time across the entire operational range.

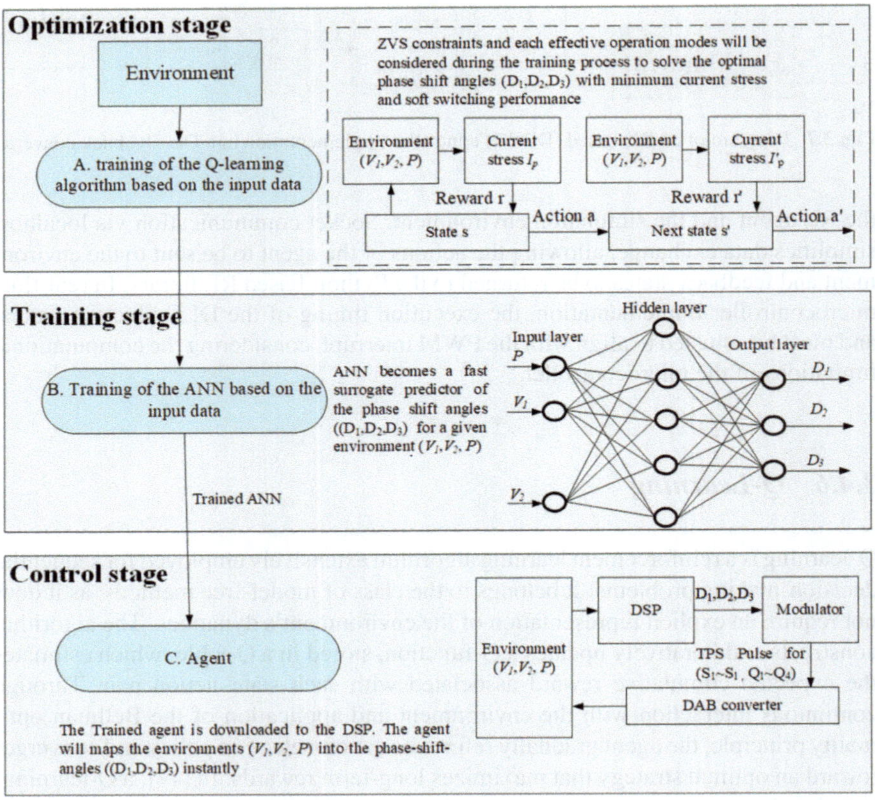

Fig. 3.8 Flow diagram of the proposed RL-ANN based minimum-current-stress scheme for a dual-active-bridge converter

3.5 Guidelines and Considerations for Implementing Machine Learning in DSPs

This section outlines strategies and guidance covering the initial stages of data collection, preprocessing, training, and model evaluation. Each stage plays a crucial role in ensuring that the ML models deliver reliable, high-performing results in DSP applications [19].

3.5.1 *Data Collection and Preprocessing*

The foremost challenge in implementing ML techniques in motor drives and power electronic systems using DSPs lies in the acquisition of high-quality data. To ensure the development of a robust and reliable model, the dataset must adequately represent the full spectrum of operating conditions. However, several obstacles complicate this process. For instance, generating experimental data for critical scenarios, such as fault detection, is not only difficult but also poses the risk of damaging the hardware. Consequently, constructing datasets that capture all possible operating conditions becomes a significant limitation. Furthermore, the number of features that can be reliably extracted from experimental setups is often restricted, further constraining the comprehensiveness of the available data. This limitation can hinder the performance of the model since important features may not be adequately captured. To address these challenges, previous studies have proposed various methods. One strategy involves creating synthetic data through mathematical modeling and augmenting it with limited experimental observations to improve precision. An alternative technique focuses on collecting experimental measurements and subsequently refining the dataset through filtering and smoothing processes to better match simulated results. Transfer learning can also be utilized to minimize large dataset requirements by applying knowledge from similar applications. Feature engineering techniques are applied to merge several parameters into unified representations, simplifying data complexity. Furthermore, interpretable ML methods assist in determining the most significant variables from extensive feature sets, thus enhancing model transparency and facilitating dimensionality reduction, as discussed in [20].

Moreover, a particularly critical challenge is handling out-of-distribution (OOD) data, which refers to inputs that differ significantly from the training data. This discrepancy can lead to unreliable predictions, since the model is unprepared for such inputs. For instance, ANNs designed to classify current sensor faults may struggle with OOD data. To mitigate this challenge, two key strategies have been implemented: introducing random noise during training and applying temperature scaling during inference. These techniques adjust the prediction confidence without adding significant computational complexity, ultimately improving model robustness when facing unexpected data that could otherwise lead to misclassification. In addition, unbalanced data arises when certain conditions, such as normal operation, are much more frequent than rare events, such as faults. This imbalance can cause ML models to be biased toward the majority class, resulting in poor detection of rare but critical events, such as faults or anomalies. As a result, models may exhibit reduced sensitivity, increasing the likelihood of false negatives, which is especially problematic in applications such as predictive maintenance and fault detection in motor drives or power converters. To address unbalanced data in power electronics, techniques such as data augmentation (generating synthetic fault data) or oversampling the minority class can help balance the dataset. Additionally, using specialized algorithms such as cost-sensitive learning or applying resampling strategies can improve model performance on rare but critical events.

3.5.2 Model Training

Selecting the appropriate ML algorithm and training method is critical for power electronics and motor drive applications in DSP systems. There are three main approaches to train an ANN which is a widely implemented algorithm in DSPs. The first is offline training, where the network is trained using algorithms such as Levenberg–Marquardt. Following the initial training phase, the optimized weights and biases of the neural network can be extracted and applied directly in real-time applications. A second strategy is to retrain the model. In this approach, the network is first trained offline and then refined using experimental data in an online environment. This process enhances robustness and improves generalization when operating conditions vary. A third strategy is fully online training, where the network parameters are updated continuously in real time through algorithms such as stochastic gradient descent or steepest descent. In addition, the extended Kalman filter is also employed to enable efficient and stable real-time training of artificial neural networks in dynamic systems.

3.5.3 Model Evaluation

Evaluating a ML model is a crucial step to ensure its robustness and reliability. It is important to assess its performance in various simulations before applying it to an experimental setup. For example, if ML code is implemented in Python or MATLAB, the optimal weights and biases should be extracted and implemented in circuit simulation software such as PSIM to verify that the same level of performance is achieved. This process should be completed before deploying the model to a DSP. Moreover, cross-validation methods such as k-fold cross-validation, conformal prediction, and explainable ML techniques are recommended during the simulation phase before real-time deployment, because they provide more interpretable and generalized performance estimates.

In classification problems, relying solely on accuracy is often inadequate. Additional metrics such as precision, recall, and F1-score should be considered to better assess the model performance, especially in cases with imbalanced datasets. Furthermore, stability analysis, such as Lyapunov stability, is important to evaluate the robustness of control systems, as proposed in various studies.

3.6 Evaluating the Computation Time of ML on DSPs

Figure 3.9 presents various models of Texas Instruments DSPs from the TMS320 family, highlighting their core types, MIPS (millions of instructions per second), and clock frequencies. DSPs can be categorized based on core architecture. The TMS320F28377D and TMS320F28379D are dual-core processors, offering high performance with 800 MIPS at 200 MHz, making them suitable for demanding real-time control applications. The TMS320F28335 and TMS320F2812 are single-core DSPs with 150 MIPS at 150 MHz, making them well-suited for applications requiring lower computational power. The TMS320F28379S is a single-core DSP with a performance of 400 MIPS at 200 MHz, which provides a balance between efficiency and power consumption. The TMS320F28075, running at 240 MIPS with 120 MHz, is another high-performance DSP optimized for motor control and power electronics. These variations in processing capabilities cater to different industrial applications, from motor drives to power converters and real-time signal processing tasks.

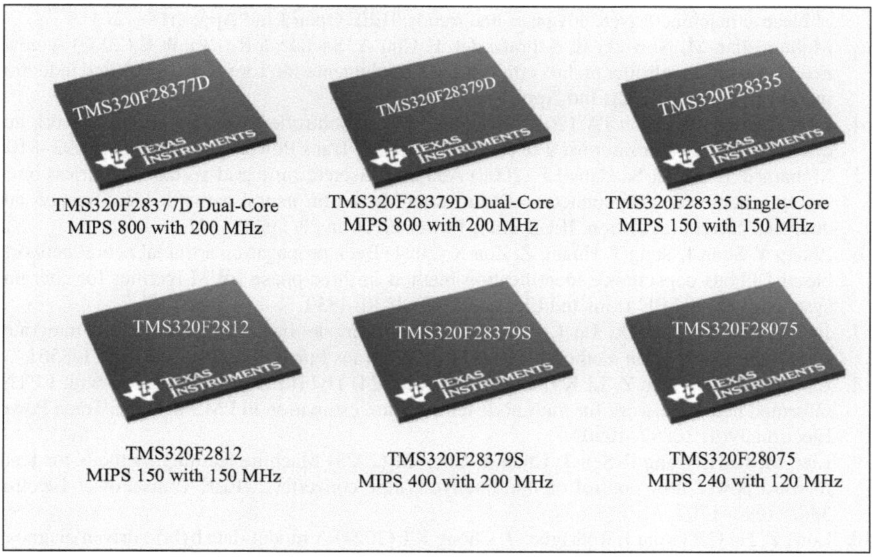

Fig. 3.9 Widely used DSP types implemented for power electronics and motor drives

Various applications of ANNs in motor control, power electronics, and fault detection have been discussed, focusing on computational time and hardware implementation [19]. For motor control, an MLP trained with simulation and experimental data for FOC of IM required 280 μs on a TMS320C30 DSP, which is significantly higher than the 7 μs required for conventional FOC implementation. In position sensorless control of PMSMs, a denoising autoencoder trained with experimental data at low speeds required a computational time of 250 μs on a TMS320F28379S DSP. This suggests that utilizing advanced DSPs or parallel processing techniques can enhance computational efficiency. For power electronics, DC bus capacitance identification in electric vehicle (EV) charging systems using an MLP achieved a computational time of 2 ms on a TMS320F28335 DSP. Similarly, ML-based methods for real-time control in power electronics often face computational constraints due to high sampling frequency requirements. For fault detection, the computational burden remains a major limitation. In T-type grid-connected converters, an MLP trained on experimental data from mathematical fault models required 2 ms on a TMS320F28379D DSP, which is excessively high, because fault detection algorithms must operate within much shorter time frames to prevent inverter damage [19].

References

1. Zhang S (2021) Artificial intelligence in electric machine drives: Advances and trends. https://arxiv.org/pdf/2110.05403.pdf
2. Zhang GT, Wallscheid O, Porrmann M (2023) Machine learning for the control and monitoring of electric machine drives: advances and trends. IEEE Open J Ind Appl 4:188–214
3. Mohamadian M, Nowicki E, Ashrafzadeh F, Chu A, Sachdeva R, Evanik E (2003) A novel neural network controller and its efficient DSP implementation for vector controlled induction motor drives. IEEE Trans Ind Appl 39(6):1622–1629
4. Tran HN, Le KM, Jeon JW (2019) Adaptive current controller based on neural network and double phase compensator for a stepper motor. IEEE Trans Power Electron 34(8):8092–8103
5. Mohamed YARI, El-Saadany EF (2008) Adaptive discrete-time grid-voltage sensorless interfacing scheme for grid-connected DG-inverters based on neural-network identification and deadbeat current regulation. IEEE Trans Power Electron 23(1):308–321
6. Zhang Y, Shan J, Song T, Huang Z, Zhu X (2024) Back propagation artificial neural network-based DC bus capacitance identification method in three-phase PWM rectifier for charging system of EVs. IEEE Trans Ind Electron 71(5):4830–4839
7. Jin L, Mao Y, Wang X, Lu L, Wang Z (2024) A model-based and data-driven integrated temperature estimation method for PMSM. IEEE Trans Power Electron 39(7):8553–8561
8. Liu W, Kong X, Fan Z, Li K, Peng R, Qu R (2024) Hybrid thermal modeling with LPTN-informed neural network for multinode temperature estimation in PMSM. IEEE Trans Power Electron 39(9):10897–10909
9. Liao M, Li H, Wang P, Sen T, Chen Y, Chen M (2023) Machine learning methods for feed-forward power flow control of multi-active-bridge converters. IEEE Trans Power Electron 38(2):1692–1707
10. Long Z, He C, Garcia J, Rodríguez J, Chong KT (2024) A model-data hybrid driven diagnosis method for open-switch faults in three-phase T-type grid-connected converters. IEEE J Emerg Sel Top Power Electron 12(4):3925–3935

11. Li Z, Gao Y, Zhang X, Wang B, Ma H (2021) A model-data-hybrid-driven diagnosis method for open-switch faults in power converters. IEEE Trans Power Electron 36(5):4965–4970
12. Lin F-J, Huang M-S, Chen S-G, Hsu C-W, Liang C-H (2020) Adaptive backstepping control for synchronous reluctance motor based on intelligent current angle control. IEEE Trans Power Electron 35(7):7465–7479
13. El-Sousy FFM, Abuhasel KA (2017) Intelligent adaptive dynamic surface control system with recurrent wavelet Elman neural networks for DSP-based induction motor servo drives. In: Proceedings of the 2017 IEEE industry applications society annual meeting. Cincinnati, OH, USA, pp 1–16
14. El-Sousy FFM (2013) Adaptive dynamic sliding-mode control system using recurrent RBFN for high-performance induction motor servo drive. IEEE Trans Ind Inform 9(4):1922–1936
15. Wang L, Zhu ZQ, Bin H, Gong L (2021) A commutation error compensation strategy for high-speed brushless DC drive based on Adaline filter. IEEE Trans Ind Electron 68(5):3728–3738
16. Kwon S, Yoon C, Lee Y-I (2022) Practical implementation method of reinforcement learning for power converter. IFAC-PapersOnLine 55(9):437–441
17. Tang Y, Zhang S, Zhu H, Zhan Y, Guo F (2021) Reinforcement learning-based efficiency optimization scheme for the DAB DC–DC converter with triple-phase-shift modulation. IEEE Trans Ind Electron 68(8):7350–7361
18. Tang Y, Zhang S, Zhan Y, Guo F, Zhao Y (2022) RL-ANN-based minimum-current-stress scheme for the dual-active-bridge converter with triple-phase-shift control. IEEE J Emerg Sel Top Power Electron 10(1):673–689
19. Al-Kaf HAG, Lee KB, Blaabjerg F (2025) Overview of DSP-based implementation of machine learning methods for power electronics and motor drives. J Power Electron 25:271–288
20. Al-Kaf HAG, Lee K-B (2023) Explainable machine learning method for open fault detection of NPC inverter using SHAP and LIME. In: Proceedings of the 2023 IEEE conference on energy conversion (CENCON). Kuching, Malaysia, pp 1–14

Chapter 4
Machine Learning Methods for Controlling PMSM Drives

Abstract This chapter presents advanced control strategies for Permanent Magnet Synchronous Motor (PMSM) drives, addressing the limitations of conventional methods such as PI controllers and Model Predictive Control (MPC) in terms of computational burden, steady-state accuracy, and transient response. By integrating machine learning (ML) techniques, particularly neural networks (NNs), the chapter proposes three innovative solutions. First, the NN-MPC-DSVPWM approach utilizes offline-trained NNs to reduce the execution time of conventional MPC with discrete space vector PWM, enabling efficient real-time implementation on DSPs without compromising performance. Second, a hybrid controller combining MPC and Field-Oriented Control (FOC) with NN-based switching methods achieves fast dynamic response, minimal steady-state error, and uniform switching through SVPWM. This method proves highly robust and effective through extensive validation. Finally, the Generalized Hybrid Method (GHM) extends control to multilevel inverters by integrating FOC, MPC, fuzzy logic (FL), and NN for rapid and intelligent decision-making. GHM effectively reduces overshoot and enhances system stability and adaptability. Together, these ML-driven strategies significantly enhance PMSM drive performance across diverse operating conditions, offering scalable and intelligent solutions for industrial automation, transportation.

4.1 Conventional Current Controls for PMSM Drives

In modern electrical drive systems, precise and efficient motor control is paramount. Two leading control strategies that have been extensively studied and implemented are FOC and MPC. This section delves into the mathematical formulations of these control techniques, exploring their underlying principles, advantages, and disadvantages, with a focus on empirical validation through experimental results.

H. A. G. Al-kaf and K.-B. Lee, *Machine Learning Integration in Power Electronics*,
https://doi.org/10.1007/978-981-95-3845-4_4

4.1.1 Field Oriented Control (FOC)

The mathematical equations for a PMSM in the d-q reference frame, as utilized in FOC, are given in (4.1).

$$\begin{cases} V_d = k_{pd} I_{derr} + k_i I_{dierr} T_s - \omega_e L_q I_q \\ V_q = k_{pq} I_{qerr} + k_i I_{qierr} T_s + \omega_e L_d I_d + \omega_e \phi_f \end{cases}, \tag{4.1}$$

where variables V_d and V_q represent the voltage components in the rotating *d-q* reference frame. The terms I_{derr}, I_{qerr}, I_{dierr}, I_{qierr}, I_d, and I_q correspond to the current errors, integration errors, and estimated currents along the d and q axes. The parameters L_d and L_q denote the inductances, ω_e indicates the rotor's electrical angular speed, and φ_f is the rotor-generated flux linkage. The proportional and integral control gains are given by k_p and k_i and their values are determined as described in Eq. (4.2).

$$\begin{cases} k_{pd} = L_d \omega_{cc} \\ k_{pq} = L_q \omega_{cc} \\ k_i = R_s \omega_{cc} \end{cases} \tag{4.2}$$

The parameter ω_{cc} represents the bandwidth of the current controller and is commonly chosen as one-tenth of the switching frequency f_{sw} [1]. The stator resistance per phase is denoted by *Rs*.

Subsequently, the voltages are converted into the stationary α-β reference frame using the inverse Park transformation. The switching signals for the two-level inverter are generated using SVPWM by comparing the ABC components in (2.1) with the PWM carrier, ensuring a fixed switching frequency. While this method achieves uniform loss distribution and lowers current ripple, the use of PI regulators in FOC restricts the speed of current response. To overcome this limitation, model predictive control (MPC) is considered a superior strategy for enabling fast and accurate current control in PMSMs. A general block diagram of FOC for a PMSM is shown in Fig. 4.1a.

4.1.2 Model Predictive Control (MPC)

MPC is an advanced control strategy widely used in industrial applications to optimize future control actions based on a system's predictive model. Figure 4.1b illustrates the block diagram of MPC, which integrates multiple future control variables into a single cost function, enabling MPC to have a fast dynamic response while maintaining a relatively simple implementation.

The mathematical model of a PMSM in the d–q reference frame can be described by the voltage equations shown in (4.3).

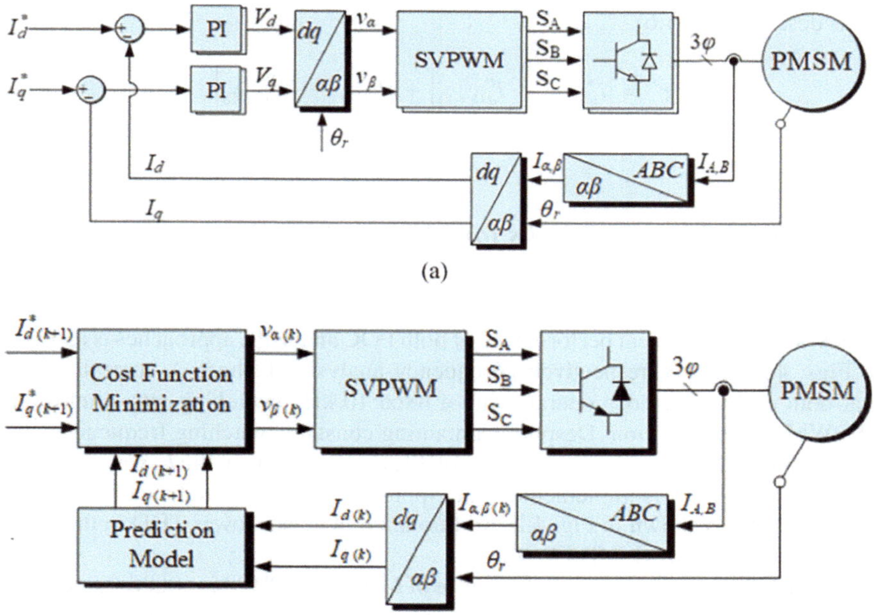

(a)

(b)

Fig. 4.1 Block diagram of conventional control methods: **a** FOC; **b** MPC-DSVPWM

$$\begin{cases} V_d = R_s I_d + L_d \dfrac{d}{dt}(I_d) - \omega_e L_q I_q \\ V_q = R_s I_q + L_q \dfrac{d}{dt}(I_q) + \omega_e L_d I_d + \omega_e \phi_f \end{cases} \tag{4.3}$$

The discrete-time representation is obtained by estimating the derivative dI_{dq}/dt using the forward Euler method, as illustrated in Eq. (4.4).

$$\frac{d}{dt}(I_{dq}) = \frac{I_{dq(k+1)} - I_{dq(k)}}{T_s}, \tag{4.4}$$

The predicted current of the PMSM can be calculated using Eq. (4.5), where T_s represents the sampling time.

$$\begin{cases} I^p_{d(k+1)} = I_{d(k)} + \dfrac{T_s}{L_d}\left(-R_s I_{d(k)} + L_q \omega_e I_{q(k)} + V_{\alpha(k)}\right) \\ I^p_{q(k+1)} = I_{q(k)} - \dfrac{T_s}{L_q}\left(R_s I_{q(k)} + L_d \omega_e I_{d(k)} + L_q \omega_e \phi_f - V_{\beta(k)}\right) \end{cases}, \tag{4.5}$$

where the voltage vectors $V_{\alpha(k)}$ and $V_{\beta(k)}$ are the optimize to minimize the cost function g^p.

as described in (4.6).

$$g^p = \left|I^*_{d(k+1)} - I^p_{d(k+1)}\right| + \left|I^*_{q(k+1)} - I^p_{q(k+1)}\right|. \tag{4.6}$$

4.1.3 Experimental Results for MPC and FOC

Table 4.1 lists the specifications of the 5 kW PMSM utilized in the study. The steady-state and transient performance of both FOC and MPC approaches is assessed in Figs. 4.2 and 4.3, respectively. Frequency analysis of phase A current reveals that both control methods operate with a fixed 10 kHz switching rate, confirming SVPWM implementation. Despite maintaining constant switching frequencies, the MPC-DSVPWM technique introduces notable current ripple in the I_q waveforms. In contrast, when FOC is implemented, the ripple is significantly reduced by approximately 82.5% as shown in Fig. 4.2. Additionally, FOC has lower THD in the phase A waveform than MPC-DSVPWM.

Nevertheless, the dynamic performance of MPC-DSVPWM is notably superior to that of FOC. Figure 4.3 illustrates the system response to a step change in reference current from 1 to 5 A. As shown in Fig. 4.3b, MPC-DSVPWM enables rapid response of the target reference with minimal oscillation. In comparison, Fig. 4.3a shows that although FOC eventually tracks the updated reference, its slower response is less suitable for industrial applications requiring rapid dynamic performance.

Table 4.1 Simulation PMSM parameters

Parameters	Value [Unit]
Rated power	5 [kW]
Moment of inertia	0.00666 [kg m^2]
Pole (P)	8
d-axis inductance (L_d)	0.00729 [H]
q-axis inductance (L_q)	0.00725 [H]
Stator resistance (R_s)	0.158 [Ω]
Magnetic flux linkage (λ_m)	0.264 [Wb]
Rated speed	1750 [r/min]
Rated torque	27.3 [Nm]
Sampling time (T_s)	100 [μs]
Rated current	17.23 [A]

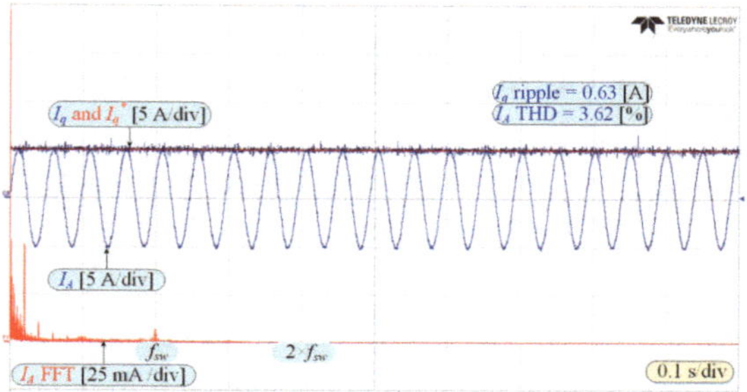

(a)

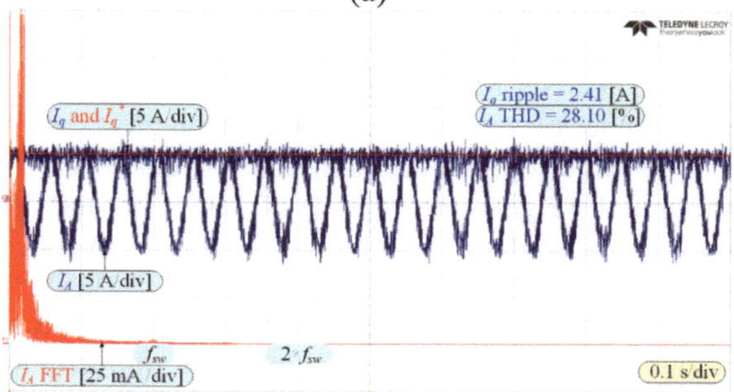

(b)

Fig. 4.2 Experimental assessment of the steady-state operation with $I_q* = 5$ A at 300 r/min: **a** field-oriented control combined with SVPWM (FOC-SVPWM); **b** model predictive control combined with DSVPWM (MPC-DSVPWM)

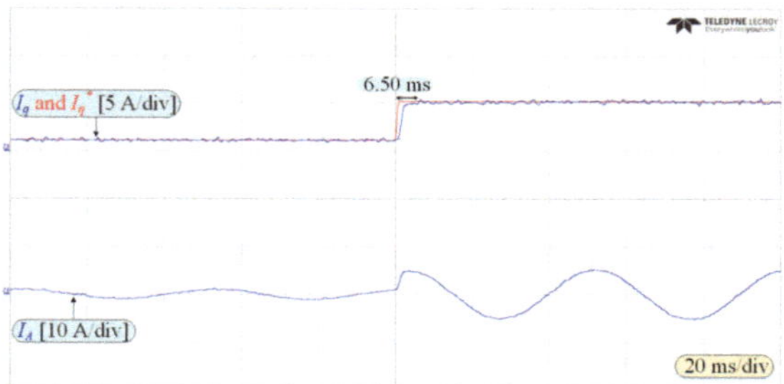

(a)

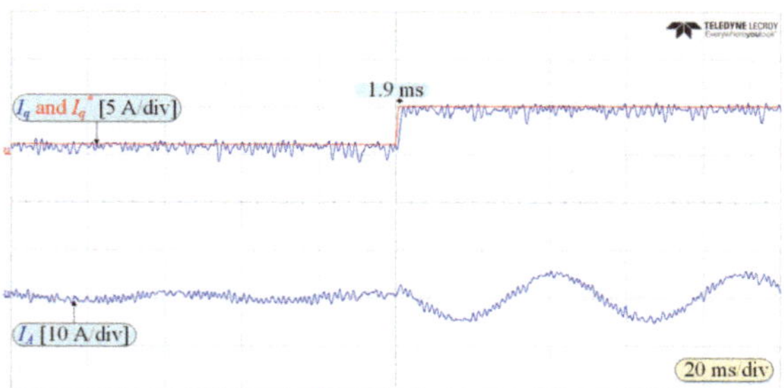

(b)

Fig. 4.3 Experimental assessment of transient-state operation with I_q^* from 1 to 5 A at 300 r/min: **a** field-oriented control combined with SVPWM (FOC-SVPWM); **b** model predictive control (MPC)

4.2 Low Complexity Hybrid ANN-MPC-DSVPWM for Emulating Multilevel Inverters

In this section, the computation burden and steady-state performance of MPC in DSPs are discussed and analyzed in detail. In addition, SVPWM integration is examined as a strategy for enhancing the steady-state performance of MPC, with attention to its effect on the associated increase in computation time. Furthermore, multilayer neural network-based MPC (MLNN-MPC) in motor drives has received significant attention for addressing the computation time challenges of MPC in power electronic applications. However, low-order DSPs are often limited by the computational burden, especially when many virtual voltage vectors are generated. The execution time of MLNN-MPC in low-order DSPs is greatly influenced by factors such as the number of hidden layers, neurons, targets, and the type of activation function. Furthermore, optimizing numerous parameters is required for MLNN, including the initial weights, the number of iterations, and the number of neurons. To overcome these limitations, single-layer neural network combined with MPC-DSVPWM (SLNN-MPC-DSVPWM). The SLNN-MPC-DSVPWM offers several advantages, including simple DSP implementation for industrial use, the capability to handle numerous VVs through switching state optimization, and the removal of lookup table initialization for all VVs. The SLNN-MPC-DSVPWM is based on the Levenberg–Marquardt algorithm, achieving a lower execution time compared with MPC-DSVPWM and MLNN-MPC. The SLNN-MPC-DSVPWM approach is validated through comprehensive simulation and experimental evaluation using PMSM drive system.

4.2.1 Conventional MPC-DSVPWM for PMSMs

Based on the mathematical model of MPC for PMSMs described in previous sections, integrating DSVPWM with MPC allows for flexible positioning of $v_{\alpha(k)}$ and $v_{\beta(k)}$ within the control region of the inverter. Figure 4.4 illustrates the control region of a two-level three-phase inverter, where the actual voltage vectors are marked with circles and the virtual vectors are indicated with squares. A total of 37 VVs are generated and uniformly distributed over three hexagons. This study employs DSVPWM due to its straightforward implementation and excellent performance. It ensures a constant switching frequency without complex calculations for each VV. Furthermore, SVPWM increases the modulation range by 15.5% over sinusoidal PWM. Its implementation on most DSPs is straightforward due to the use of a predefined ePWM function [1]. Three-phase reference voltages v^*_A, v^*_B, and v^*_C are obtained from $v_{\alpha(k)}$ and $v_{\beta(k)}$ using inverse transformation techniques. Based on DSVPWM principles, the offset voltage v^*_{sn} is calculated as detailed in Eq. (4.7).

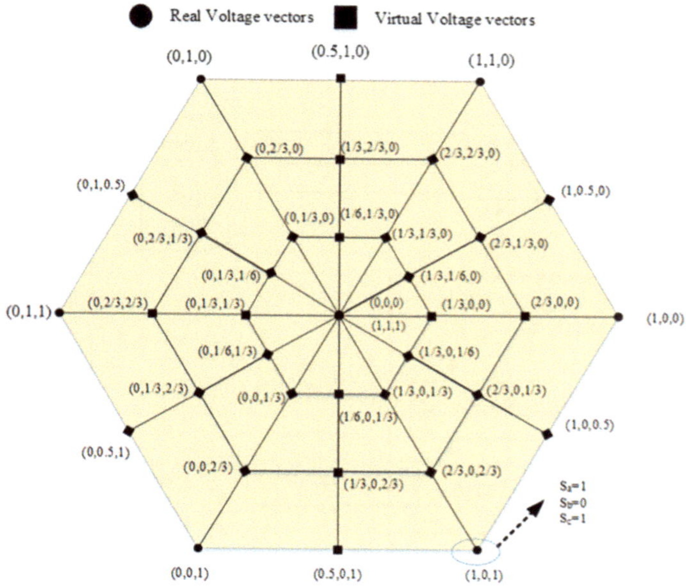

Fig. 4.4 Space vector representation using 37 VVs

$$V_{sn}^{*} = \frac{-\left(V_{\max}^{*} + V_{\min}^{*}\right)}{T_s}, \tag{4.7}$$

Among the reference phase voltages V^*_A, V^*_B, and V^*_C, the largest and smallest values are denoted as $V^*_{\max}$ and $V^*_{\min}$ [1]. For ease of computation, the reference pole voltage of phase V^*_{An}, is obtained using the relation in (4.8). The same approach can be extended to phases B and C. In this formulation, V^*_{sn} is incorporated into V^*_A, as given in (4.8).

$$V_{An}^{*} = V_{A}^{*} + V_{sn}^{*}. \tag{4.8}$$

From (4.9), the normalized phase-A reference is derived.

$$V_{Az}^{*} = \frac{\left(2 + V_{An}^{*}\right)}{V_{dc}}, \tag{4.9}$$

where V_{dc} indicates the DC-link voltage. Switching signal generation is accomplished by comparing V^*_{Az} to a PWM triangular carrier V_{CA}, as detailed in Eq. (4.10).

$$\begin{cases} V_{Az}^{*} \geq V_{CA}, & S_A = 1; \quad \overline{S_A} = 0 \\ V_{Az}^{*} < V_{CA}, & S_A = 0; \quad \overline{S_A} = 1 \end{cases}, \tag{4.10}$$

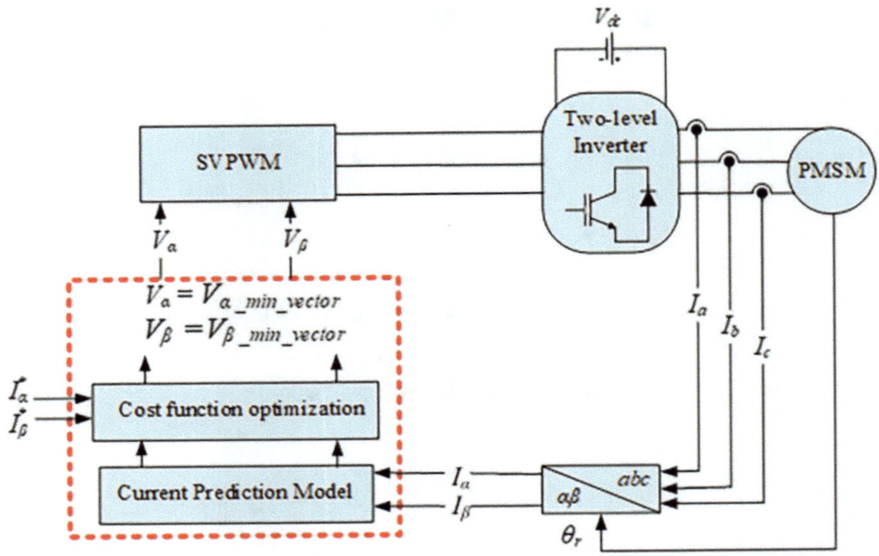

Fig. 4.5 Block diagram of model predictive control combined with discrete SVPWM (MPC-DSVPWM)

S_A and $\bar{S}_A$ denote the inverter switching signals. Figure 4.5 presents the comprehensive block diagram of conventional MPC-DSVPWM.

4.2.2 *ANN-MPC-DSVPWM for PMSMs*

As discussed in previous studies, implementing MPC-DSVPWM and MLNN-MPC-DSVPWM in real-time is challenging due to computational time constraints. Additionally, an SLNN trained with the Levenberg–Marquardt algorithm has demonstrated superior predictive accuracy compared to advanced neural networks, such as MLNN and extreme learning machines (ELMs), when applied to specific complex multivariable problems [2]. Consequently, this study introduces single-layer neural network with DSVPWM to address the computational challenges associated with MLNN-MPC and traditional MPC-DSVPWM, particularly.

In PMSM drive systems implemented on low-order DSPs, managing a large set of VVs can be computationally demanding. To address this, the SLNN uses a dual-activation strategy: a computationally function such as the sigmoid is applied during the training phase, while a simpler function such as the linear form is adopted during real-time DSP operation, ensuring efficiency without loss of control accuracy. Moreover, the SLNN eliminates the dependency on predefined VV look-up tables by directly assigning appropriate values to the switching signals and employing mathematical formulations within DSVPWM to generate the required voltage vectors.

It is important to emphasize that the SLNN method was not developed to exceed the performance of conventional MPC-DSVPWM and MLNN controllers. The key benefit of the SLNN approach lies in its execution time, which remains unaffected by system complexity. Furthermore, it achieves the fastest execution among all methods implemented on DSPs.

The proposed SLNN-MPC-DSVPWM general block diagrams are shown in Figs. 4.6 and 4.7 [3]. SLNN offline training substitutes for the computationally demanding current prediction model and cost function evaluation requirements. Data for SLNN development under the MPC framework is generated through PSIM simulations. The input set includes I_{α}, I_{β}, I^*_{α}, and I^*_{β} in the stationary frame, while the outputs correspond to the switching states S_A, S_B, *and* S_C. The dataset is divided into 70% for model training and 30% for validation using MATLAB. Training is carried out offline to reduce the heavy computational burden of real-time learning. The proposed scheme employs three independent SLNNs, each responsible for selecting the appropriate switching state for a phase. Using the expressions of V_{α} and V_{β} in (4.11) [1], both real and virtual voltage vectors can be synthesized without additional initialization. For example, when the SLNN determines the optimal state as $S_A = 1$, $S_B = 0$, and $S_C = 0$, substitution into (4.11) gives $V_{\alpha} = 2/3V_{DC}$ and $V_{\beta} = 0$, corresponding to V_1.

$$
\begin{aligned}
v_{\alpha} &= \frac{V_{dc}}{3}(2S_A - S_B - S_C),\\
v_{\beta} &= \frac{V_{dc}}{\sqrt{3}}(S_B - S_C).
\end{aligned}
\tag{4.11}
$$

Subsequently, V_{α} and V_{β} are mapped to V_A, V_B, and V_C using an inverse transformation. Then they are applied to the inverter via DSVPWM. The use of three SLNNs eliminates the need for initializing a lookup table for numerous virtual VVs. The SLNN-MPC-DSVPWM execution time is reduced by eliminating the computational burden associated with large multiplication matrices between inputs and all available VVs. Instead, the method simplifies computations to basic multiplications between inputs and corresponding phase switch states. This enhancement not only improves MPC performance but also enables implementation with numerous VVs, unlike MPC-DSVPWM and MLNN, which face significant execution time constraints. The SLNN process is depicted in Fig. 4.8.

The SLNN is mathematically expressed in Eq. (4.12).

$$
\begin{aligned}
H &= g(x),\\
x &= WI + b,
\end{aligned}
\tag{4.12}
$$

where H is the output vector of the output layer, I is the input vector of the network, b is the bias vector, W are the weight values that needs to optimized, and g is a sigmoid function, which can be expressed by (4.13).

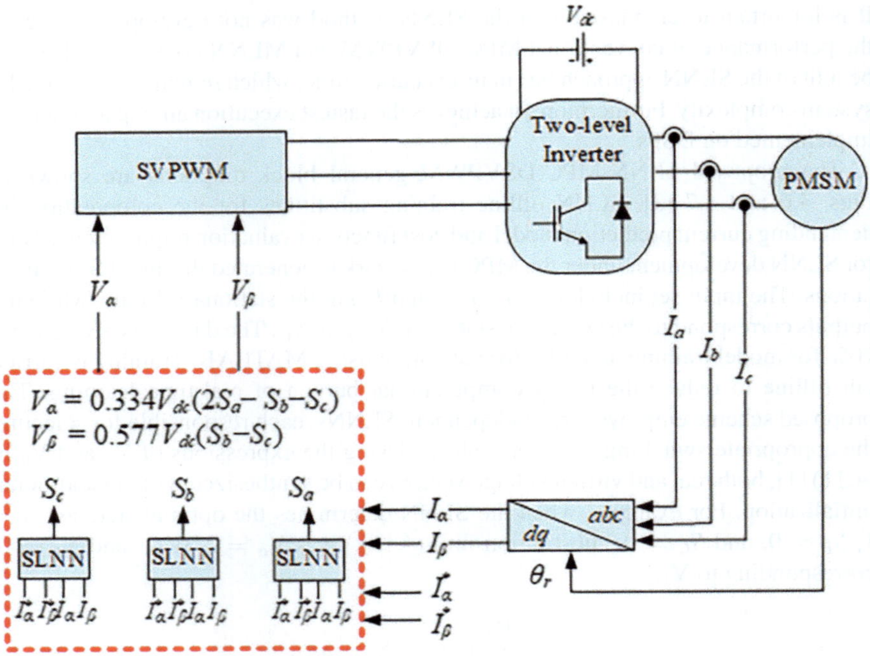

Fig. 4.6 Block diagram of a single layer neural network combined with the MPC-DSVPWM control method (SLNN-MPC-DSVPWM)

$$g = \frac{1}{1 + e^{-x}}. \tag{4.13}$$

DSP-based sigmoid function implementation suffers from extended execution times. For this reason, the SLNN approach utilizes sigmoid functions only in the training process. Fast sigmoid and linear activation functions achieve faster execution, as indicated in Eq. (4.14).

$$\begin{aligned} g &= \frac{1}{1 + |x|}, \\ g &= x. \end{aligned} \tag{4.14}$$

The SLNN architecture for training and testing phases is depicted in Fig. 4.9. Training employs sigmoid functions to establish accurate input–output relationships. Testing utilizes optimal *W* and *b* parameters with linear activation functions to achieve minimal execution time. Superior algorithms enable SLNN improvements that exceed MLNN and ELM performance. The Levenberg–Marquardt algorithm manages weight updates, as shown in Eq. (4.15).

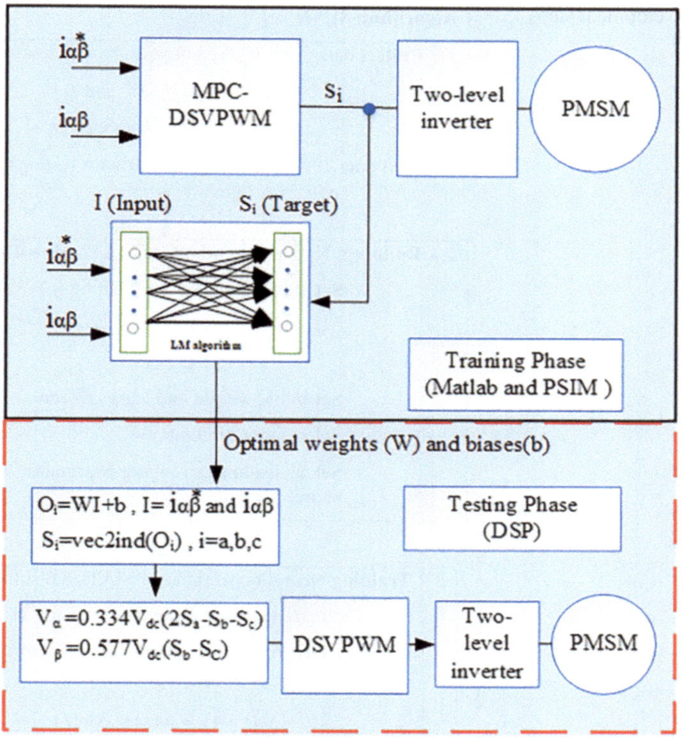

Fig. 4.7 Training and testing steps of a single-layer neural network combined with the MPC-DSVPWM control method (SLNN-MPC-DSVPWM)

$$
\begin{aligned}
W(t+1) &= W(t) - \Delta W(t), \\
\Delta W(t) &= \left(JMJM^{t} + \mu I\right)^{-1} JM^{t} er,
\end{aligned}
\tag{4.15}
$$

where JM is the Jacobian matrix, er represents the error term, $W(t+1)$ indicates the weight adjustments at each iteration, I is the identity matrix, and μ is the combination coefficient.

A comparison of the SLNN's computational load with that of the MLNN introduced in [4] is presented in (4.16).

$$
\begin{aligned}
N_{SLNN} &= HI, \\
N_{MLNN} &= 21g + 2gH + ag + bH,
\end{aligned}
\tag{4.16}
$$

where I is the input layer, H is the output layer, g is the hidden layer, and a and b are the number of calculations in the activation function. The SLNN parameters were set for 10 iterations. Performance analysis revealed that increasing iterations beyond

Fig. 4.8 Development steps of the SLNN

Algorithm: SLNN

1 **Collect** data using PSIM software

Input $I = [I_{\alpha}, I_{\beta}, I_{\alpha}^{*}, \text{and } I_{\beta}]$

Target $H = [S_x \text{ where } x \text{ is } a,b,c]$

Split data 70% for training and 30% for testing data

2 **Building** Neural network using MATLAB software

Set sigmoid activation function (g)

$$g = \frac{1}{1 + e^{-x}}$$

Set initial weight and biases to zeros

Set number of iteration

Set algorithm Levenberg-Marquardt for training neural network

3 **Training** Neural network using MATLAB software

Estimate the error using mean square error

Update Weight (w) and biases (b) factors according to Levenberg-Marquardt

$$\begin{cases} w(t+1) = w(t) - \Delta w(t) \\ \Delta w(t) = (JJ^{t} + \mu I)^{-1} J^{t} e \end{cases}$$

4 **Testing Neural network in real-time DSP**

Collect the optimal weight and biases

Change the sigmoid activation function to the linear activation function

Classify and choose the optimal S_x

10 yielded no further enhancement. Therefore, 10 was selected as the optimal value. The dataset is partitioned into 70% for training and 30% for testing, as described in Table 4.2. PSIM simulation-based performance assessment using various current and speed values is essential to validate that the SLNN learns system behavior rather than merely memorizing data. The training phase requires an extensive dataset that is uniformly distributed across broad current and speed ranges. Consequently, data collection exceeded one million points under diverse operational scenarios.

A rigorous training process is crucial to prevent both overfitting and underfitting. The SLNN-MPC achieves high accuracy in overall performance for switching state

(a)

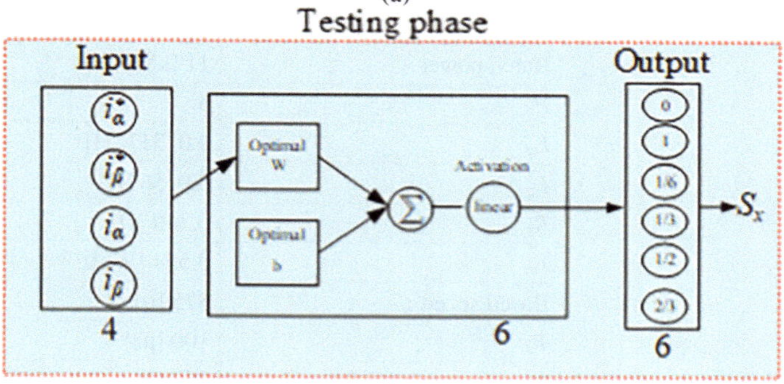

(b)

Fig. 4.9 Structure of an SLNN: **a** during the training stage; **b** during the testing stage

Table 4.2 Dataset distribution by percentage

Dataset distribution	Percent (%)
Training dataset	70
Testing dataset	30

A, achieving approximately 80% accuracy. Switching states B and C also demonstrated similar accuracy levels. MPC-DSVPWM commonly relies on observers to handle parameter variations in motor-drive applications. SLNN-MPC-DSVPWM overcomes this limitation by incorporating additional input variables that enable observer-based parameter correction.

4.2.3 Simulation Validation

The proposed SLNN-MPC effectiveness is analyzed and validated through PSIM-based simulation results in this section. Simulation parameters are detailed in Table 4.3. Performance comparison was conducted between the proposed approach and MPC-DSVPWM. Without hardware execution time restrictions, simulations employed 12, 25, and 37 VVs. Table 4.4 lists the performances of each activation function in the testing phase. It presents the execution time and THD for the A-phase current, as well as the rise time of the I_q current. Among the tested activation functions, the linear function yielded the shortest execution time while still providing superior performance compared to both the sigmoid and fast-sigmoid functions. With 37 voltage vectors, the linear function required only about 8 μs, whereas the sigmoid and fast-sigmoid functions took around 80 and 40 μs, respectively. Nevertheless, during the training phase, the sigmoid activation proved effective in capturing the nonlinear input–output mapping of MPC-DSVPWM. Figure 4.10

Table 4.3 Simulation PMSM parameters

Parameters	Value [Unit]
Rated power	11 [kW]
P	6
L_d	0.01317 [H]
L_q	0.0156 [H]
R_s	0.349 [Ω]
λ_m	0.554 [Wb]
Based speed	875 [rpm]
T_s	100 [μs]
V_{DC}	300 [V]

presents the steady-state A-phase current waveforms obtained using conventional MPC-DSVPWM and the proposed SLNN-MPC-DSVPWM. The speed was 150 rpm and the reference current was 10 A. The MPC-DSVPWM performed better than SLNN-MPC-DSVPWM at different numbers of VVs. Meanwhile, SLNN-MPC-DSVPWM with 25 and 37 VVs exhibited better performance than MPC-DSVPWM with 12 VVs. In addition, a clear improvement was observed when the number of VVs was increased. At 150 rpm, the reference current was increased from 2 to 10 A to evaluate the dynamic response performance. Both methods demonstrated similar results for the I_{q} current waveform at 12 and 25 VVs, as shown in Fig. 4.11. However, both methods had a slower dynamic response when 37 VVs were used. In addition, a clear reduction in the I_q current ripple was observed when the number of VVs was increased. Figure 4.12 compares the performance of conventional MPC-DSVPWM and SLNN-MPC-DSVPWM across different motor speeds. The speed reference ranged from 0 to 875 rpm, corresponding to the base speed, based on the applied DC-link voltage. As shown, both methods accurately track the reference I_q^* current. Moreover, the SLNN-MPC-DSVPWM demonstrates reduced I_{q} ripple and produces a nearly sinusoidal A-phase current waveform. As a result, SLNN-MPC-DSVPWM successfully learned the behavior of MPC-DSVPWM under various operational conditions, as verified by the simulation results.

4.2.4 Experimental Validation

Experimental validation confirmed the advantages of the proposed SLNN-MPC-DSVPWM compared to conventional MPC-DSVPWM and recent MLNN-MPC-DSVPWM approaches. Figure 4.13 shows the experimental configuration of a two-level inverter and low-order DSP TMS320F28335. Machine loading was performed using an induction motor controlled by a commercial YASKAWA inverter. Due to hardware constraints, execution time evaluation of MPC-DSVPWM was conducted without full system operation. The algorithm was implemented on DSP hardware

Table 4.4 Performance comparison of activation functions

Activation function	Number of VVs	Execution time (μs)	I_A THD (%)	I_q rise time (ms)
Sigmoid	12	30	5.6	0.9
Fast sigmoid		20	5.6	0.9
Linear		7	5.6	0.9
Sigmoid	25	50	2.9	1.8
Fast sigmoid		25	2.9	0.9
Linear		7	2.87	0.9
Sigmoid	37	80	2.1	1.6
Fast sigmoid		40	2	1.2
Linear		8	2.1	1.17

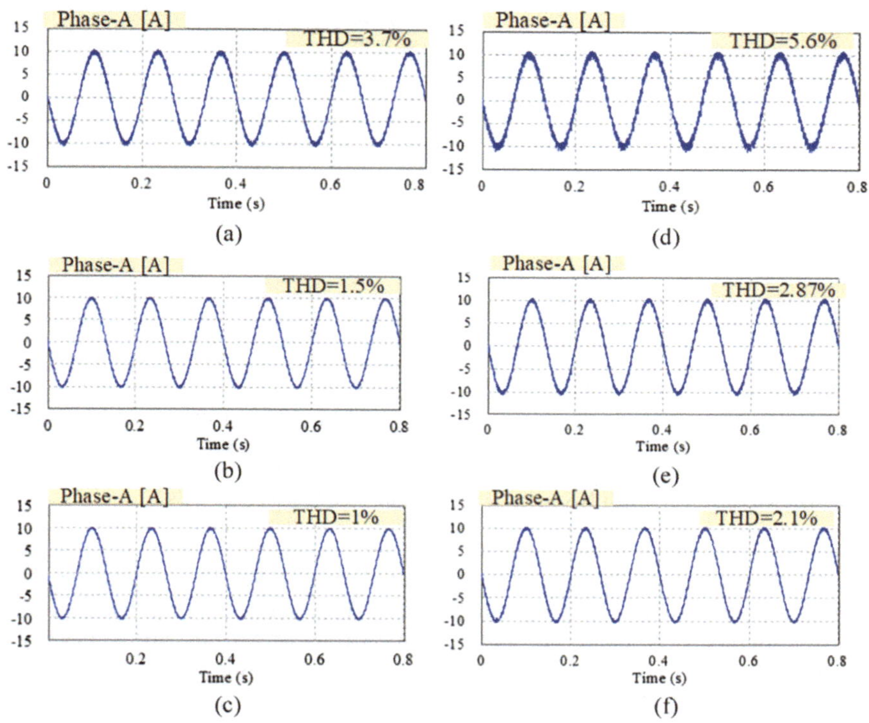

Fig. 4.10 Simulation assessment of A-phase current waveforms at various numbers of VVs: **a** MPC-DSVPWM method based on 12 VVs; **b** MPC-DSVPWM method based on 25 VVs; **c** MPC-DSVPWM method based on 37 VVs; **d** SLNN-MPC-DSVPWM method based on 12 VVs; **e** SLNN-MPC-DSVPWM method based on 25 VVs; **f** SLNN-MPC-DSVPWM method based on 37 VVs

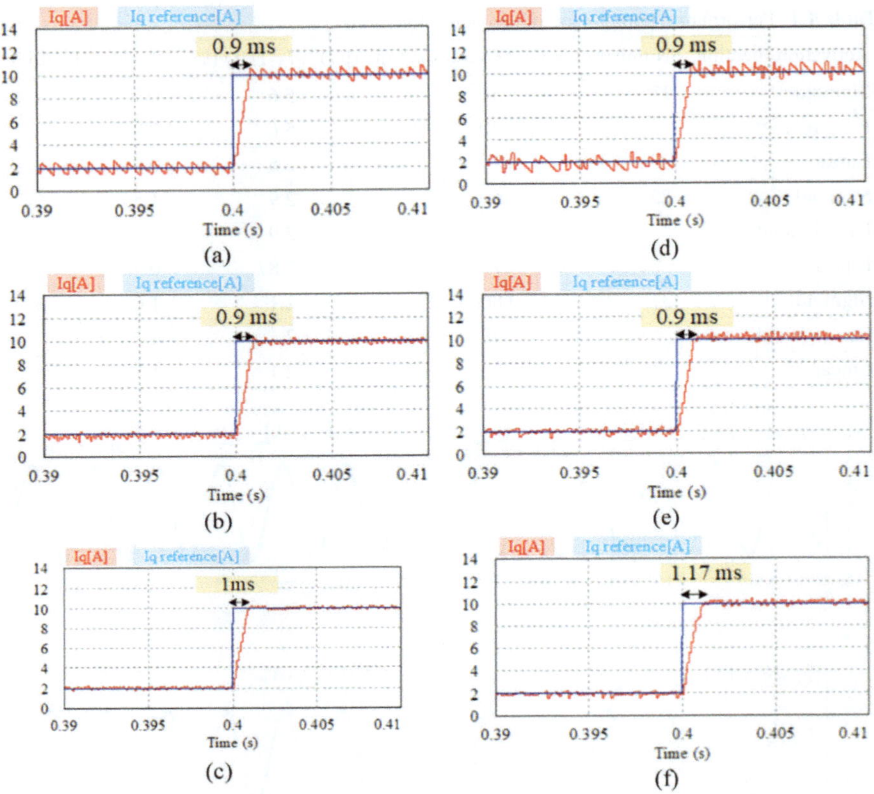

Fig. 4.11 Simulation assessment of Iq in transient at various numbers of VVs: **a** MPC-DSVPWM method based on 12 VVs; **b** MPC-DSVPWM method based on 25 VVs; **c** MPC-DSVPWM method based on 37 VVs; **d** SLNN-MPC-DSVPWM method based on 12 VVs; **e** SLNN-MPC-DSVPWM method based on 25 VVs; **f** SLNN-MPC-DSVPWM method based on 37 VVs

to measure execution times across varying numbers of VVs. The proposed SLNN-MPC-DSVPWM successfully operated with 37 VVs due to its minimal execution time requirements. Notably, the high execution time of MLNN-MPC-DSVPWM and MPC-DSVPWM for large VV values limits their applicability, making the proposed SLNN-MPC-DSVPWM the only practical option for 37 VV implementation. Figures 4.14 and 4.15 show experimental results of I_q, the A-phase current waveform, THD, and FFT. MPC-DSVPWM and SLNN-MPC- DSVPWM with 12 VVs have similar results with THD values of 28.8 and 30%, respectively. SLNN-MPC-DSVPWM with 25 and 37 VVs shows a better performance than MPC-DSVPWM with 12 VVs. A clear reduction in the THD for SLNN-MPC-DSVPWM is shown when the number of VVs is increased. The THD for SLNN-MPC-DSVPWM with 12, 25, and 37 VVs are 30, 15.5, and 10%, respectively. Both methods have a constant switching frequency of 10 kHz with the SVPWM method.

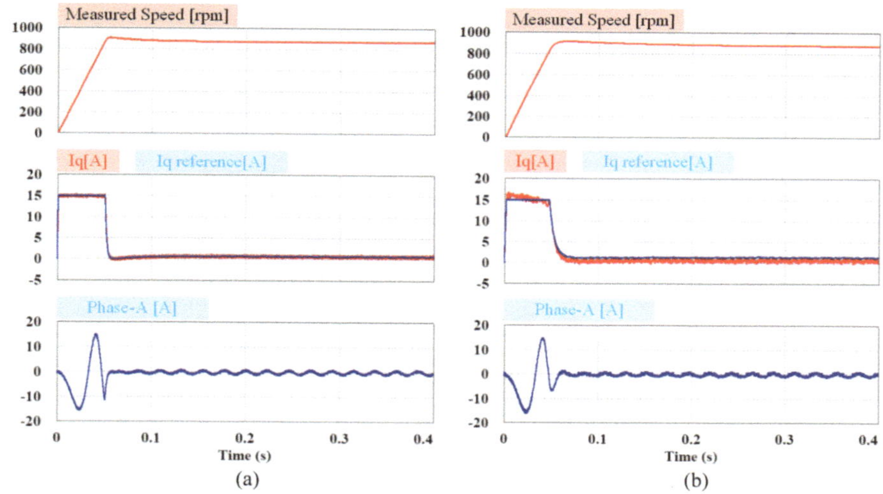

Fig. 4.12 Simulation results under a no-load test for speed change from 0 to 875 rpm: **a** model predictive control combined with DSVPWM (MPC-DSVPWM) using 37 VVs; **b** proposed single-layer neural network (SLNN-MPC-DSVPWM) using 37 VVs

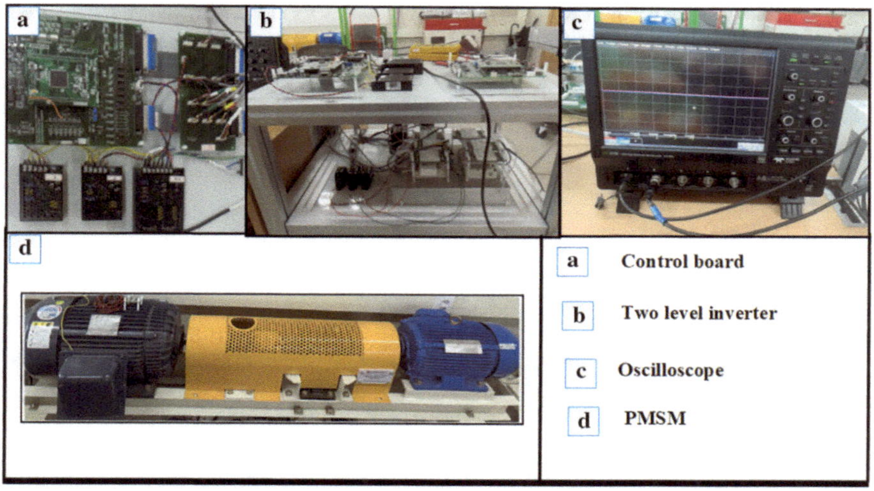

Fig. 4.13 Experimental setup

Due to the high execution time of MPC-DSVPWM, only SLNN-MPC-DSVPWM was tested over 37 VVs for motor speed variation, as shown in Fig. 4.16. It can be observed that the I_q current waveform perfectly follows the I_q reference.

Figure 4.17 presents the execution time of the MLNN-MPC when deployed on a TMS320F28335 DSP. For an MLNN configuration of 8-42-6, where 8 represents

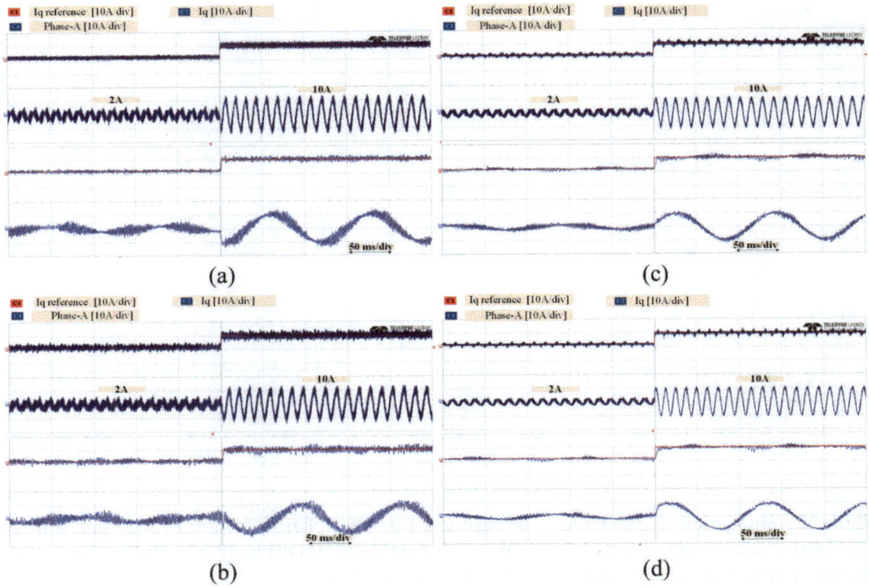

Fig. 4.14 Experimental assessment of I_q current for conventional and proposed methods: **a** MPC-DSVPWM with VVs = 12; **b** SLNN-MPC-DSVPWM with VVs = 12; **c** SLNN-MPC-DSVPWM with VVs = 25; **d** SLNN-MPC-DSVPWM with VVs = 37

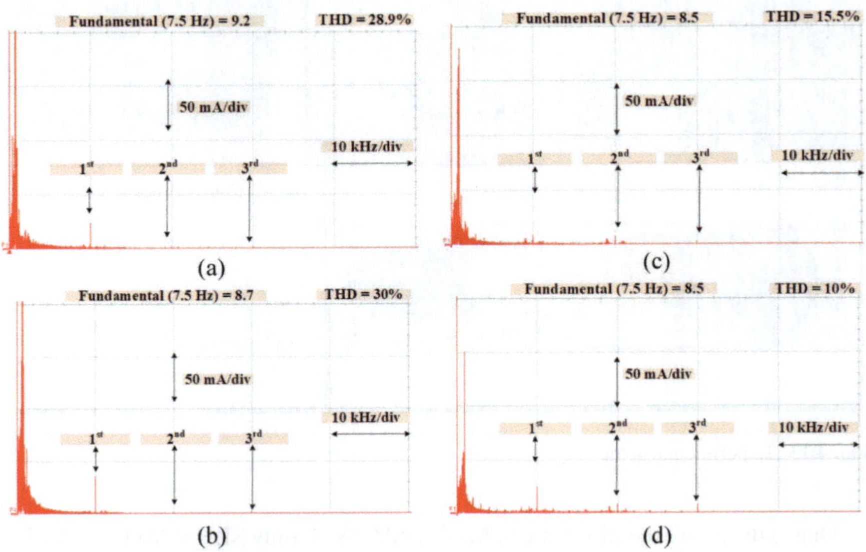

Fig. 4.15 Experimental THD performance of A-phase current: **a** MPC-DSVPWM with VVs = 12; **b** SLNN-MPC- DSVPWM with VVs = 12; **c** SLNN-MPC-DSVPWM with VVs = 25; **d** SLNN-MPC-DSVPWM with VVs = 37

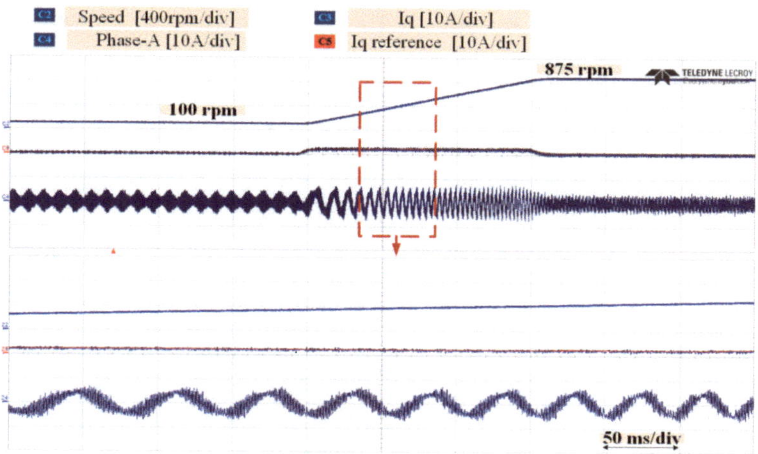

Fig. 4.16 Experimental outcomes of SLNN-MPC-DSVPWM during speed changes from 100 to 875 rpm under no-load conditions

the input nodes, 42 is the number of hidden layer neurons, and 6 corresponds to the outputs, the measured execution time is 250 μs. The hidden layer uses the tanh activation function, while the output layer employs the sigmoid function. As illustrated in Fig. 4.17a [4], the execution time is significantly influenced by the size of the hidden layer. In [4], 42 neurons were utilized in the hidden layer, necessitating extensive matrix multiplication and the computation of sigmoid functions, resulting in prolonged execution times, especially when employing a low-order DSP and serial execution. Adopting parallel execution methods is essential to achieve an execution time of less than 100 μs. Additionally, Two additional MLNN structures, specifically 4/15/7 and 8/15/7, were utilized. Their execution times are shown in Fig. 4.17b, c. The findings indicate that deploying MLNN-MPC on low-end DSPs becomes impractical as the hidden layer neuron count increases. Furthermore, increasing the number of VVs also prolongs the execution time, since each VV corresponds directly to an output node in the network. However, the proposed method reduces execution time significantly by optimizing switching states in place of VVs.

Figure 4.18 presents the variation in computation time for each control method as the number of voltage vectors changes. For instance, SLNN-MPC-DSVPWM achieves an execution time of only 7 μs with 12 VVs, compared to 30 μs required by MPC-DSVPWM. Furthermore, the proposed method completes all calculations within a maximum of 8 μs with 37 VVs, while conventional MPC-DSVPWM takes 75 μs under the same conditions. SLNN-MPC-DSVPWM achieves lower execution times primarily because it eliminates the need for lookup table initialization, which is required in conventional MPC-DSVPWM. The adoption of a linear activation function further decreases computation time compared with sigmoid or fast-sigmoid functions. In addition, this approach allows for longer prediction horizons, improving control performance without a significant increase in computational

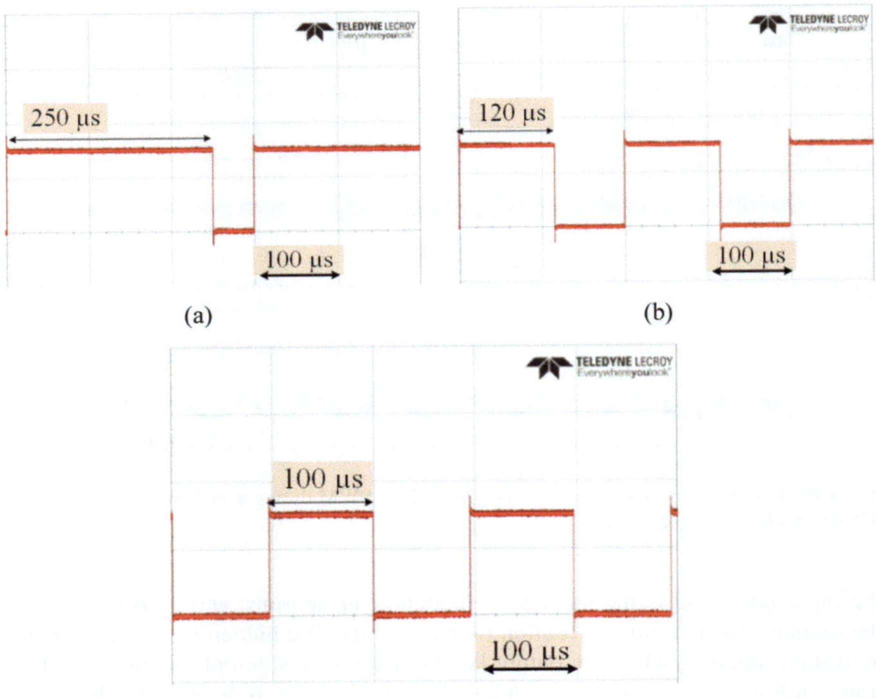

Fig. 4.17 Experimental execution time of different MLNN structures for MLNN-MPC: **a** 8 inputs, 42 hidden neurons, 6 outputs; **b** 8 inputs, 15 hidden neurons, 7 outputs; **c** 4 inputs, 15 hidden neurons, 7 outputs

effort. In contrast, MPC-DSVPWM exhibits exponential growth in computation time with increasing step prediction horizons, driven by the exponential rise in possible switching states.

4.3 Hybrid ANN-Based Current Control Using Robust Switching Methods

After analyzing the results of SLNN-MPC-DSVPWM in Sect. 4.3, it is evident that SLNN achieves comparable performance to MPC while enhancing its steady-state performance. However, despite these improvements, the steady-state performance remains inadequate. This section introduces an FOC-MPC current controller for a PMSM supplied by a two-level inverter, combining the rapid response of MPC with the steady-state accuracy of FOC. SVPWM maintains a constant switching frequency and streamlines the generation of switching signals for the inverter. Two proposed

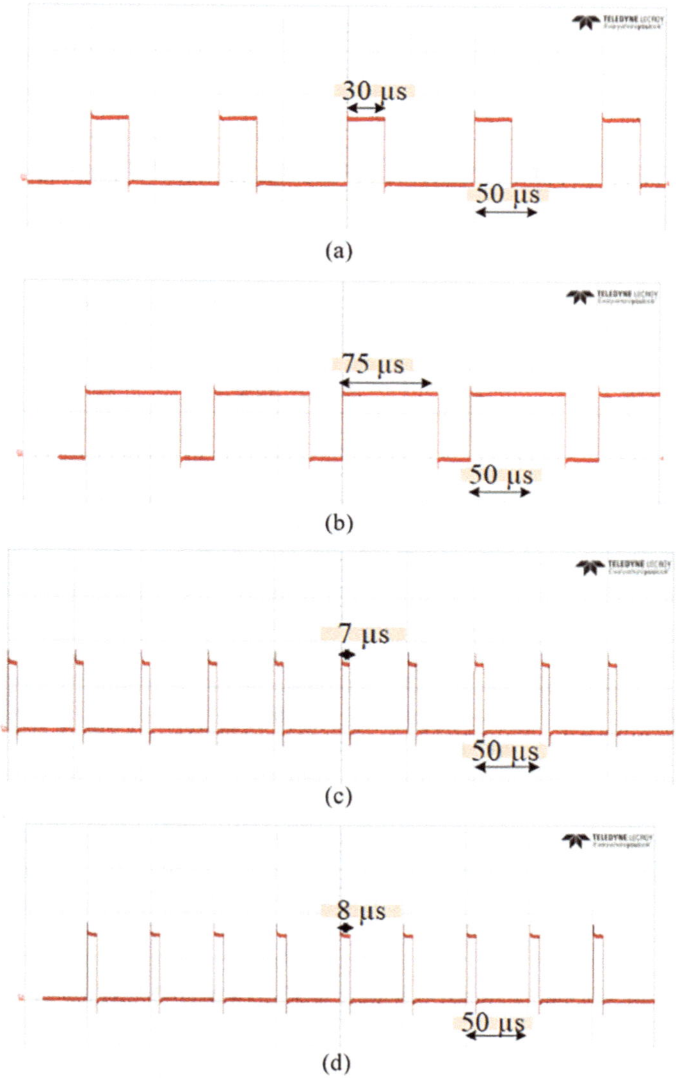

Fig. 4.18 Experimental execution time for conventional and proposed methods: **a** MPC-DSVPWM with VVs = 12; **b** MPC-DSVPWM with VVs = 37; **c** SLNN-MPC-DSVPWM with VVs = 12; **d** SLNN-MPC-DSVPWM with VVs = 37

switching Methods (PSMs) are employed to ensure smooth transitions between FOC and MPC, preventing unwanted oscillations or spikes. These PSMs utilize hysteresis, average current ripple, and artificial neural networks. They are evaluated against two conventional switching methods (CSMs), with simulation and experimental results demonstrating the superior performance of PSMs. The results confirm that PSMs enable smooth and robust switching between MPC and FOC, deliver faster dynamic response, and maintain zero steady-state error under various operating conditions.

4.3.1 Conventional Hybrid Control and Its Limitations

The constant k_t (CSM-I), formulated in (4.17), quantifies the ratio between I_q and I^*_q in percentage terms. A value of $k_t = 100\%$, indicates that the current error is completely eliminated. The appropriate controller is selected based on the range of k_t specified in (4.17).

$$k_t = \left(1 - \frac{I_q^* - I_q(k)}{I_q^*}\right) \times 100 \begin{cases} \text{FOC}, \ (95 \le k_t \le 105) \\ \text{MPC}, \ (k_t < 95 || k_t > 105) \end{cases}. \tag{4.17}$$

FOC controller activation occurs when ΔIq stays within a 5% range. When this condition is not met, MPC rapidly adjusts the states to approach their reference values. A key limitation of this approach is its reliance on a fixed k_t value, which may not be suitable for all operating conditions. For instance, in low-speed experiments, ΔI_q can exceed 5%, potentially causing the MPC method to be unnecessarily selected during steady-state operation. Moreover, dynamic response deteriorates when k_t values drop below 5%, due to FOC activation before reaching the target. Therefore, the k_t switching approach exhibits insufficient robustness in certain system conditions, particularly in PMSM drive implementations.

The second conventional method, the hysteresis switch (HS), is also referred to as CSM-II, offers improved performance over comparators by reducing switching losses and enhancing control stability. The HS design structure are illustrated in Fig. 4.19. The configuration includes two operational bands: high (J_h,) and low (J_l). Two output states define the HS functionality. HS selection principles are presented in Eq. (4.18).

$$\begin{cases} \text{Field Oriented Control (FOC)}, \ \textit{if HS} = 2 \ \& \ \& \ \Delta I_q \le J_l \\ HS{=}HS \\ \text{Model predictive control (MPC)}, \ \textit{if HS} = 1 \ \& \ \& \ \Delta I_q \ge J_h \\ HS{=}HS \end{cases} \tag{4.18}$$

When $\Delta I_q \ge J_h$, state "2" is activated by the controller, and MPC is employed to rapidly bring the states closer to their reference values. Conversely, when $\Delta I_q \le J_l$, the proximity to the operating point triggers a switch from MPC to FOC. To avoid unnecessary controller commutations, the control scheme from k-1 is retained, mitigating the effect of noisy measurements. However, the HS has inherent drawbacks. The lower and higher bounds need to be set through testing, which can change

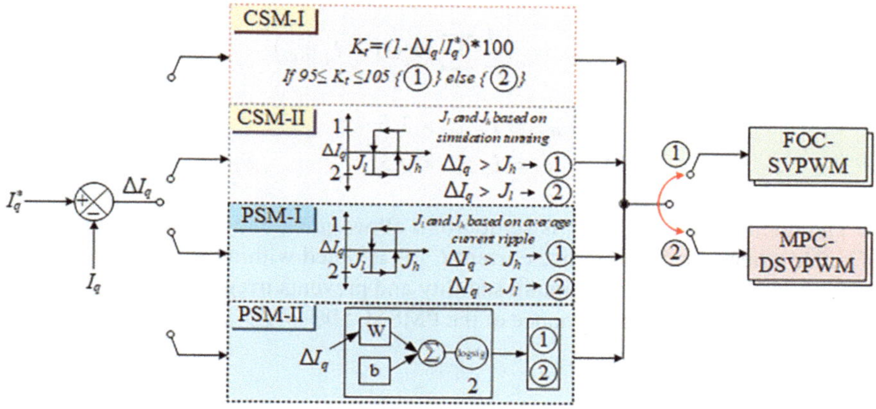

Fig. 4.19 Block diagram of the conventional and PSMs hybrid control strategy

depending on the test setup. Also, a fixed band does not work well for all operating conditions. An excessively large J_h can result in slow dynamics, whereas an overly small J_h can lead to instability, particularly during steady-state operation, resulting in undesirable bumps and oscillations. Therefore, optimal J_h value determination requires intensive fine-tuning. For instance, in an FC converter, J_h is typically chosen to be larger than the maximum voltage deviation of the converter. In motor drive applications, however, determining a suitable J_h is more complex, as it must account for multiple factors, including the operating speed and the V_{DC} signals.

CSMs suffer from a fundamental weakness in their fixed band structure, resulting in undesirable fluctuations during steady-state conditions and reduced dynamic response. Although modifying these parameters can enhance performance at specific operating points, this approach increases system complexity when specific band configurations are needed for each point. Consequently, PSM-I is implemented to provide steady-state stability through FOC and fast dynamics through MPC. Proper high and low band selection is essential, as shown in Eq. (4.19).

$$\begin{cases} \text{Field Oriented Control (FOC)}, \ \textit{if } HS = 2 \ \& \ \& \ \Delta I_q \leq J_{l\_rip.} \\ HS{=}HS \\ \text{model predictive control (MPC)}, \ \textit{if } HS = 1 \ \& \ \& \ \Delta I_q \geq J_{h\_rip.} \\ HS{=}HS \end{cases} \tag{4.19}$$

where $J_{h\_rip.}$ and J_{l_rip} are denote the high and low bands, calculated according to the Iq ripple characteristics, as expressed in Eq. (4.20).

$$I_{q\_rip.} = \sqrt{\frac{1}{r}\sum_{i=1}^{n}\left(I_q(i) - I_{q\_ave.}\right)^2}$$
$$\begin{cases} J_{h\_rip.} = I_{q\_rip.} \times 1.5 \\ J_{l\_rip.} = I_{q\_rip.} \times 0.5 \end{cases} \quad (4.20)$$

The number of samples, r, is set to 40 to allow quick and accurate computation of ripple whenever the reference current I^*_q is updated within the sampling interval T_s. This choice also helps maintain stability and prevents irregular switching during steady-state operation performance of the PMSM, 50% of I_{q_rip} is added to J_{h_rip} and subtracted from J_{l_rip}.

4.3.2 Hybrid ANN-Based Methods

The comprehensive ANN structure, referred to as PSM-II, is shown in Fig. 4.20. In addition, Fig. 4.21 provides a flowchart outlining the PSM-II switching approach. The ANN was designed to trigger FOC activation during steady-state performance and initiate MPC switching during transient events. PSM-II can therefore be characterized as an intelligent switching solution [5]. Compared to PSM-I, PSM-II eliminates the need for complex computations, providing a clear advantage.

The initial step in designing any ANN model involves gathering and processing the necessary data and splitting data between training and testing datasets. The second step is to construct the ANN model, utilizing the sigmoid function g, initializing the weight matrix w and bias b vector to zero, and setting the number of iterations to 10. The ANN model employs the Levenberg–Marquardt algorithm for updating w and b parameters, providing rapid convergence and enhanced accuracy relative to scaled conjugate gradient approaches, as shown in [2]. The third phase involves ANN training with performance assessment through accuracy measurements and confusion matrix analysis, evaluating discrepancies between actual and predicted results. Parameter optimization continues through Levenberg–Marquardt updates of w and b until minimal error is achieved. The resulting optimal w and b enable ANN classification of FOC and MPC operations following Fig. 4.21. PSM-II operates as follows. When the ANN selects "1," MPC is applied. When the ANN selects "0," the response depends on the previous state. If the previous state was "0," FOC is applied. If the previous state was "1" and ΔI_q is less than 0, MPC is applied; otherwise, FOC is selected. PSM-II is designed to ensure that I_q rapidly follows I^*_q, utilizing the benefits of FOC during steady-state operation across different system setups. The mathematical formulation of SLNN is detailed in Sect. 4.3, Eqs. (4.12)–(4.15). SLNN was trained iteratively, offline, using MATLAB. The inputs included I_q and I^*_q in the rotating reference frame, with target outputs designated as "0" for steady-state and "1" for transient-state. Experimental data were collected using an oscilloscope. The SLNN training involved 10 iterations, since increasing the number

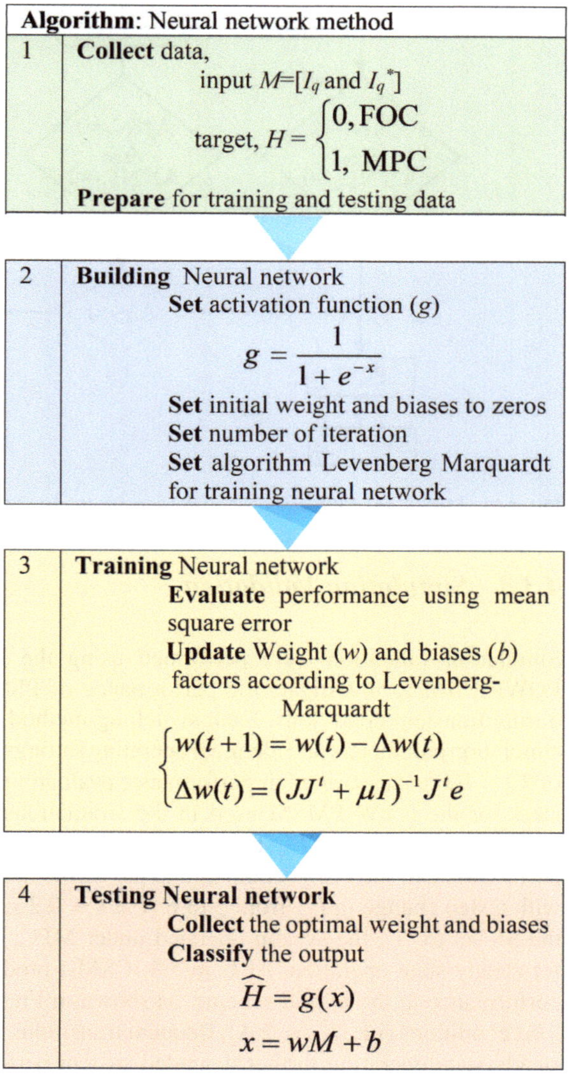

Fig. 4.20 Block diagram of the development steps for the ANN of the proposed PSM-II

of iterations beyond this did not yield additional improvements. Therefore, 10 was deemed optimal. The dataset was partitioned into 70% training data and 30% testing data.

The dataset must be extensive, comprising tens or hundreds of thousands of data points, uniformly distributed, and covering various values of I_q, I^*_q, V_{DC}, and rotation speeds.

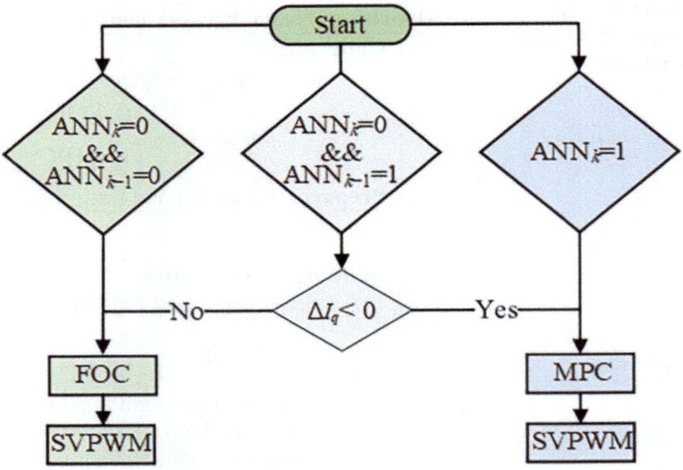

Fig. 4.21 Flowchart of the proposed switching mechanism of PSM-II

4.3.3 Simulation Validation

Simulation validation were performed using the PSIM platform developed by POWERSIM to investigate the performance of PSMs in comparison with CSMs during transient conditions. Each switching method was assessed under consistent simulation parameters and identical operating settings, employing a sampling interval of $T_s = 100$ μs, ensuring fair performance evaluation. Table 4.1 provides the parameters for the 5 kW PMSM used in the simulations. Figures 4.22 and 4.23 show simulation results of I_q, I^*_q, the switch signal (SS), and I_A. These results evaluate the performance of the PSMs compared to the CSMs at a rotational speed of 600 r/min, with a step change in I^*_q from 2 to 6 A at $t = 0.2$ s, and a DC-link of 200 V. With the SS set to '1', the system operated under MPC; otherwise, it switched to FOC for steady-state operation. At $k_t = 5\%$, CSM-I produced unstable and undesirable performance in the hybrid scheme, as shown in Figs. 4.22 and 4.23a. Under light-load conditions (i.e., $I^*_q = 2$ A), frequent transitions between FOC and MPC during steady-state operation further degraded system performance and undesired bumps and oscillations occur. These issues degrade the I_A waveform, leading to a high THD and increased I_q ripples. However, increasing the k_t value improves the steady-state performance, as shown in Figs. 4.22 and 4.23b. However, a slow-dynamic response of I_q was observed because FOC was used in situations where MPC was more appropriate. Figures 4.22 and 4.23c shows the performance of CSM-II when J_h is set to 2 A.

The value was determined using an average band configuration, where J_h should represent 10–15% of the PMSM rated current. However, this fixed band approach was not optimal across all operating conditions. Large J_h values can produce slow dynamic response, as demonstrated in Fig. 4.22d, where CSM-II failed to transition

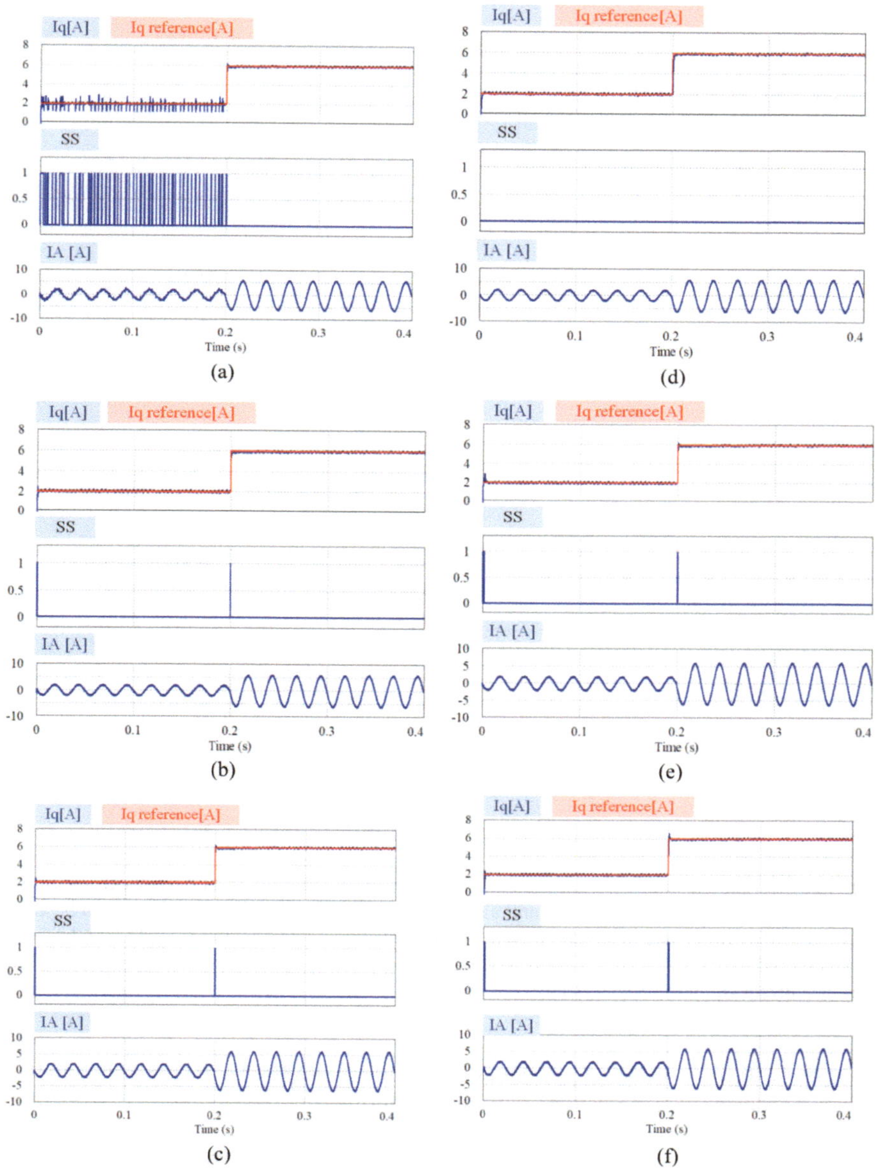

Fig. 4.22 Simulation assessment of the proposed PSMs against CSMs where I^*_q from 2 to 6 A at $t = 0.2$ s, with a V_{DC} of 200 V and at a rotational speed of 600 r/min: **a** based on fixed value (CSM-I) using $k_t = 5\%$; **b** based on fixed value (CSM-I) using $k_t = 15\%$; **c** based on hysteresis (CSM-II) method using $J_h = 2$ A; **d** based on hysteresis (CSM-II) using $J_h = 4$ A; **e** average current ripples (PSM-I); **f** based on ANN (PSM-II)

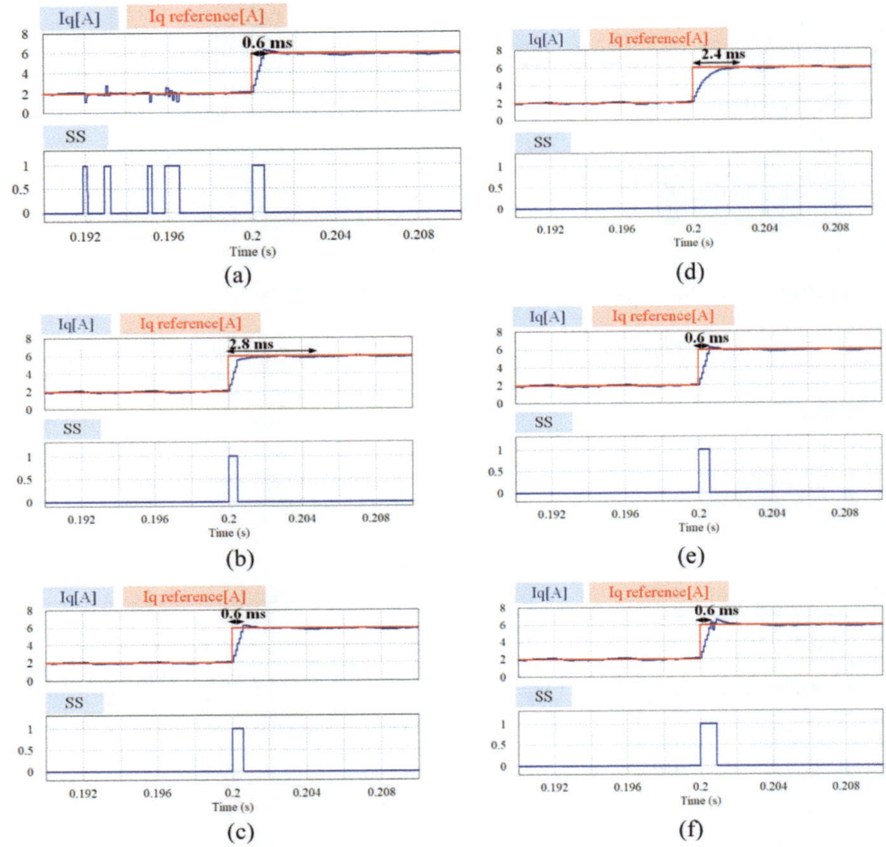

Fig. 4.23 Magnified simulation assessment of the proposed PSMs against CSMs where I^*_q is 2 to 6 A at $t = 0.2$ s, with a V_{DC} of 200 V and a rotational speed of 600 r/min: **a** conventional based on fixed value (CSM-I) using $k_t = 5\%$; **b** based on fixed value (CSM-I) using $k_t = 15\%$; **c** based on hysteresis (CSM-II) using $J_h = 2$ A; **d** based on hysteresis (CSM-II) using $J_h = 4$ A; **e** based on average current ripples (PSM-I); **f** based on ANN (PSM-II)

from FOC to MPC during I_q dynamic operation. Conversely, excessively small J_h values can destabilize the system, particularly during steady-state operation, resulting in unwanted disturbances and oscillations.

Figures 4.22 and 4.23e, f illustrate the performances of PSM-I and PSM-II. In PSM-I, FOC is applied during steady-state operation, while MPC is employed for the transient state. Its improved robustness over CSMs stems from a variable upper band value, optimized according to the average I_q ripple. This design enables fast I_q dynamics without oscillations across varying operating conditions. In PSM-II, the system effectively learns and distinguishes between the behaviors of MPC and FOC. The ANN is trained to set the state to "0" during steady-state operation and to

"1" during dynamic operation. This classification enables PSM-II to achieve a fast response that is similar to that of PSM-I.

4.3.4 Experimental Validation

The experimental configuration depicted in Fig. 4.24 confirms the effectiveness of the PSMs. It consists of a 5 kW PMSM mechanically coupled to an induction motor, controlled via a three-phase A1000 drive from YASKAWA Electric Corporation. The specifications of the PMSM, T_s, and f_{sw} are identical to those used in the simulations.

Three different system setups were utilized to assess the robustness of PSMs compared to CSMs. To ensure a fair comparison, all methods were evaluated using identical parameters across all three cases. The three cases are as follows.

(1) Experimental case A: $\omega = 300$ r/min, $V_{DC} = 100$ V, and $I^*_q = (1$ and 5 A).
(2) Experimental case B: $\omega = 600$ r/min, $V_{DC} = 200$ V, and $I^*_q = (1$ and 8 A).
(3) Experimental case C: $\omega = 1000$ r/min, $V_{DC} = 300$ V, and $I^*_q = (2$ and 10 A).

In Case A, the experimental plots of I_q, I^*_q, SS, and I_A are depicted in Fig. 4.25a. The performance of CSM-I under these conditions is unsatisfactory. Its constant k_t value leads to delayed transient behavior and unstable changes between FOC and MPC modes. Moreover, the I_A response suffers from significant ripple and waveform distortion. These drawbacks indicate that CSM-I is not a promising option for practical motor-drive implementations.

Fig. 4.24 Experimental setup

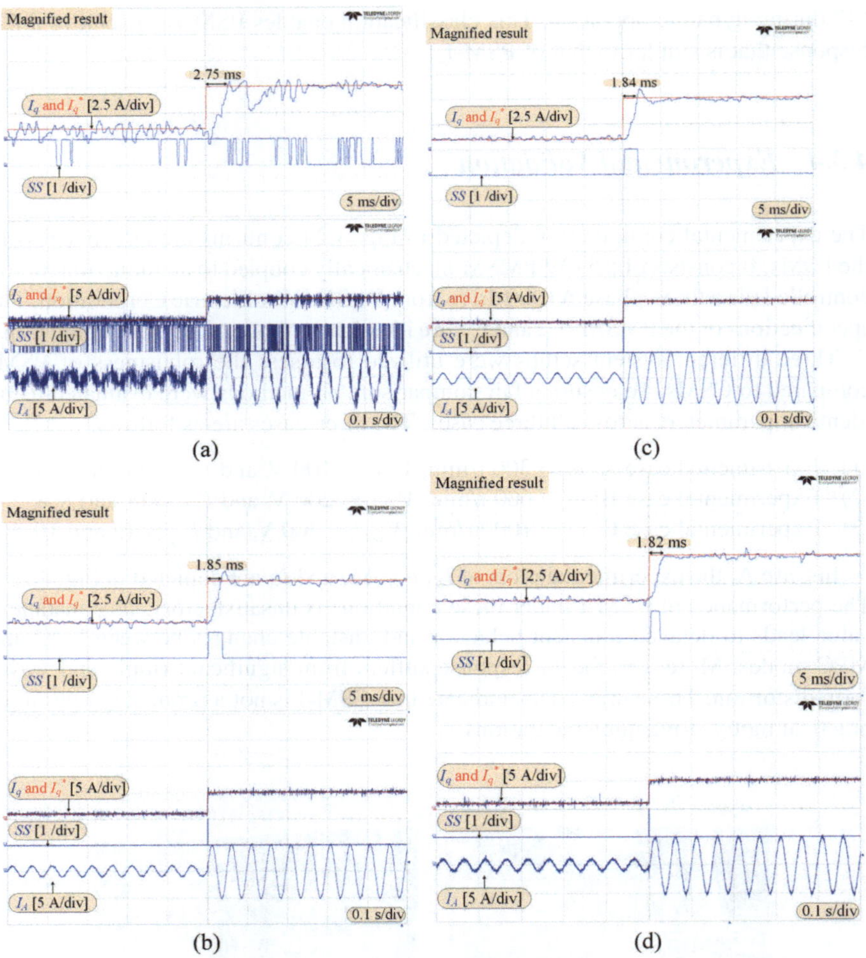

Fig. 4.25 Experimental results between the proposed PSMs and CSMs where I^*_q is 1 to 5 A and 300 r/min (Case A): **a** based on fixed value (CSM-I); **b** based on hysteresis (CSM-II); **c** based on average current ripples (PSM-I); **d** based on ANN (PSM-II)

Figure 4.25b, c, d illustrate the outcomes for CSM-II, PSM-I, and PSM-II, respectively. All three techniques enable a smooth transition from FOC to MPC during dynamic conditions. Among them, CSM-II demonstrates a comparatively longer rise time than the PSM approaches. Notably, PSM-II achieves minimal overshoot, outperforming the other methods in this aspect. A similar behavior is observed in Case B, as shown in Fig. 4.26, where PSMs deliver a quicker dynamic response than CSM-II. CSM-I was excluded from this evaluation due to its inadequate performance in Case A, which could pose risks to experimental implementation in other scenarios. Although PSM-II responds slightly slower than PSM-I, it ensures stable

FOC–MPC switching with negligible overshoot. Finally, based on Fig. 4.27, PSM-II demonstrates superior dynamic characteristics with rapid settling time and reduced overshoot during Test Case C, exhibiting better performance compared to CSM-II and PSM-I approaches. The enhanced operational behavior results from the precise ANN implementation within PSM-II and the adaptive band control mechanism integrated in PSM-I. Table 4.5 consolidates the findings obtained from all three test scenarios. The efficiency and I_A THD for all the control methods evaluated are illustrated in Table 4.6. As can be seen from this table that CSM-II and PSMs have significantly lower THD in I_A and higher efficiency compared to MPC and CSM-I for Case A at I^*_q of 5 A. The enhanced performance stems from PSM reliability, delivering rapid transient behavior matching MPC characteristics while maintaining superior steady-state operation equivalent to FOC performance.

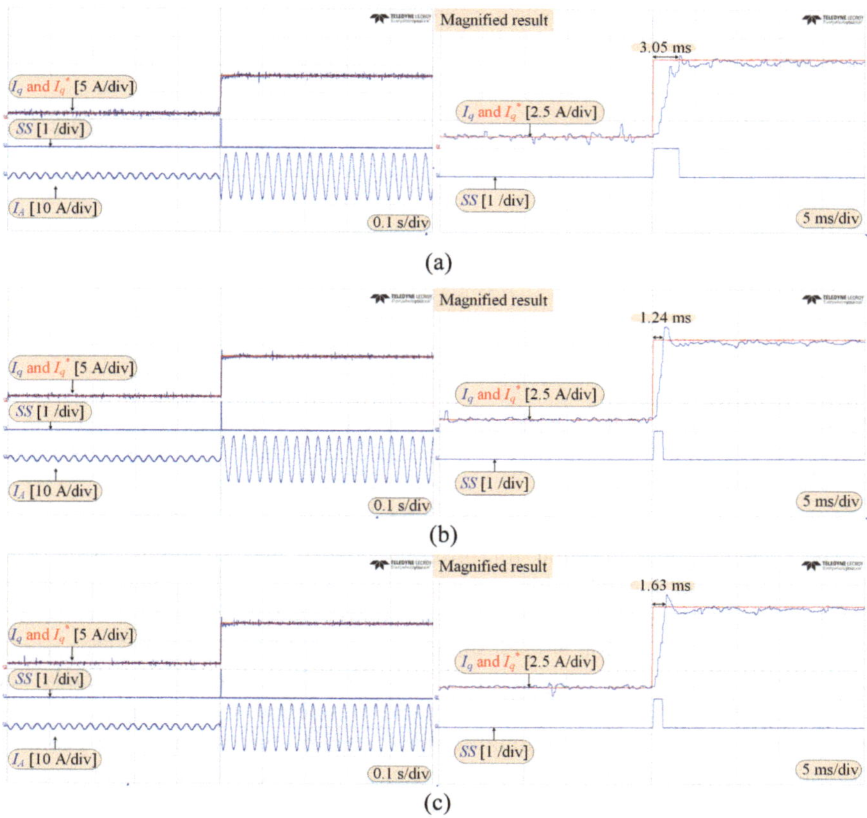

Fig. 4.26 Experimental results between the proposed PSMs and CSMs where I^*_q is 1 to 8 A at 600 r/min (Case B): **a** based on hysteresis (CSM-II); **b** based on average current ripples (PSM-I); **c** based on ANN (PSM-II)

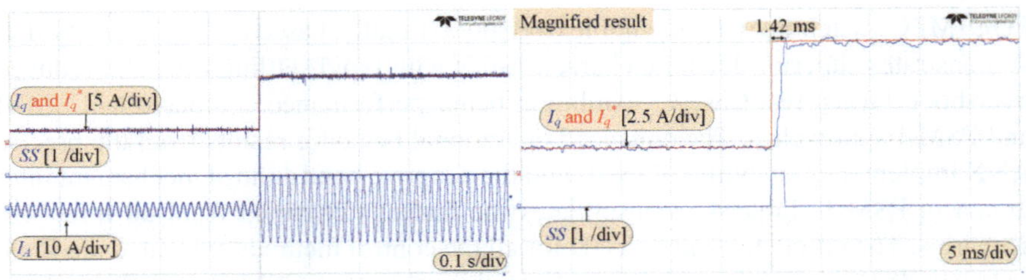

(a)

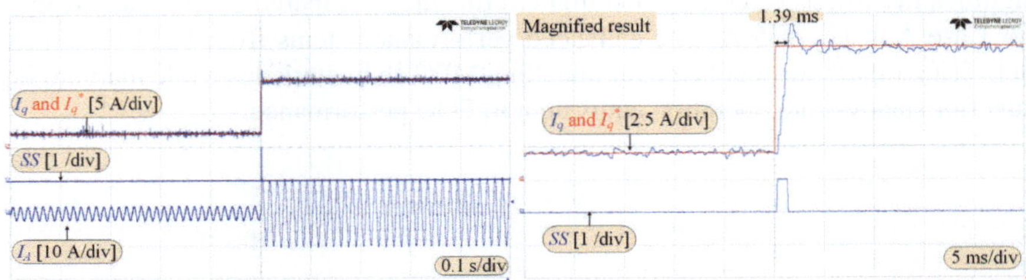

(b)

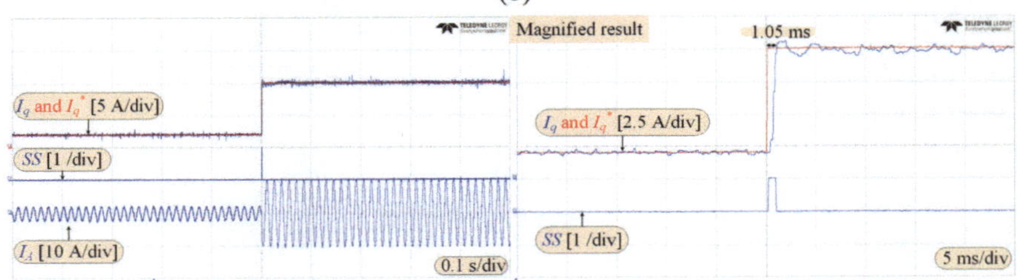

(c)

Fig. 4.27 Experimental results comparing the proposed PSMs and CSMs where I^*_q is 2 to 10 A and 1000 r/min (Case C); **a** based on hysteresis (CSM-II); **b** based on average current ripple (PSM-I); **c** based on ANN (PSM-II)

The computational duration for different control and switching methods is depicted in Fig. 4.28. MPC exhibits the longest execution time because it involves intensive computations and an iterative procedure to select the voltage vector that minimizes the cost function at each sampling instant. PSM-II also requires significant computational time compared to CSMs and PSMs. This occurs mainly due to matrix operations and the sigmoid activation function. Conversely, alternative switching techniques require minimal computational resources. The highest computational complexity occurs when PSM-II is implemented. Nevertheless, this remains manageable when implemented on industrial microcontroller applications.

Table 4.5 Comparison of experimental results for PSMs and CSMs

Switching method	Switching mechanism	Tuning required	Case	Switching between controllers	Robustness	Dynamic performance	Rise time [ms]	Overshoot [ms]
CSM-I	Fixed value	Yes	A	Failed	No	Slow	2.75	0.50
CSM-II	Fixed hysteresis band	Yes	A	Succeed	Yes	Fast	1.85	0.80
			B	Succeed	No	Slow	3.05	0.30
			C	Succeed	Yes	Fast	1.42	0.20
PSM-I	Variable hysteresis band with average current ripple	No	A	Succeed	Yes	Fast	1.84	0.75
			B	Succeed	Yes	Fast	1.24	1.00
			C	Succeed	Yes	Fast	1.39	1.25
PSM-II	Intelligent hysteresis controller	No	A	Succeed	Yes	Fast	1.82	0.001
			B	Succeed	Yes	Fast	1.63	0.90
			C	Succeed	Yes	Fast	1.05	0.50

Table 4.6 Comparison of results at 5 A for Case A

Switching method	I_a THD [%]	η [%]
Conventional FOC	3.62	86
Conventional MPC	28.10	75
Conventional CSM-I	26.34	79
Conventional CSM-II	3.83	84
Proposed PSM-I	3.76	85
Proposed PSM-II	3.80	83

4.4 MPC-Fuzzy-ANN Methods for Multilevel Inverters

This section presents an enhanced version of the previous hybrid approach (4.4), thereby creating a generalized control strategy that is applicable to multilevel inverters in PMSM systems and to eliminating the overshoot of the conventional hybrid method. The GHM leverages the benefits of field oriented control during stable operation and the capabilities of model predictive control in dynamic conditions. It targets several goals such as rapid transient response, elimination of steady-state error, minimal overshoot, reduced computation time, and require fewer VVs while avoiding the complexity of adjusting weighting parameters in multilevel inverters. To accomplish these goals, GHM integrates a single voltage vector from FOC with MPC. Additionally, fuzzy logic (FL) identifies the best VV and uses its input–output data to develop an ANN, allowing for quick and direct optimal value selection. Through experimental comparisons with traditional MPC, FOC, and modern CHMs under various operating scenarios, GHM exhibited efficient processing time, rapid

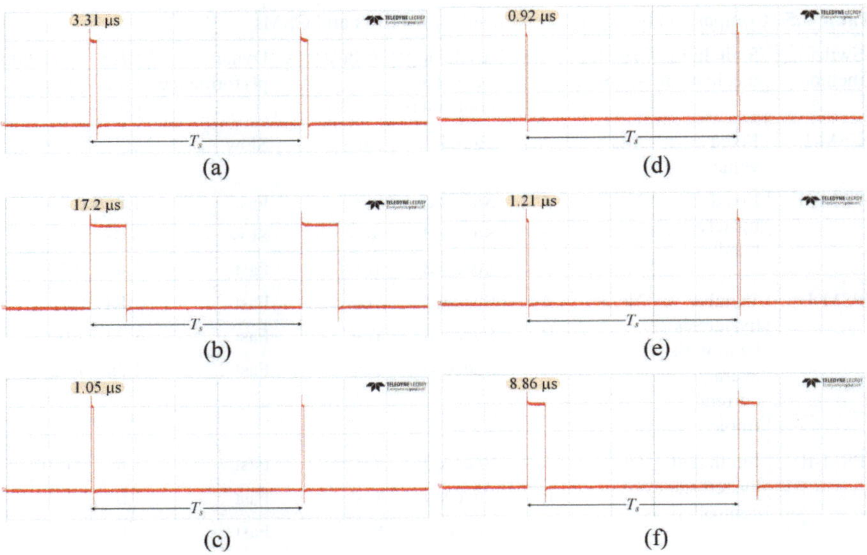

Fig. 4.28 Execution time of the proposed and conventional methods: **a** FOC; **b** MPC; **c** based on fixed value (CSM-I); **d** based on hysteresis (CSM-II); **e** based on average current ripples (PSM-I); **f** based on ANN (PSM-II)

dynamic behavior, and reliable steady-state characteristics. In contrast, recent CHMs exhibited significant overshoot issues.

4.4.1 Limitations of Conventional Hybrid Current Methods

In this section, CSM-II (from Sect. 4.3) is referred to as CHM-I. In addition, the average current ripple PSM-I (from Sect. 4.3) is referred to as CHM-II, which is based on average current ripple hysteresis control. Figure 4.29 illustrates that large overshoots can arise at certain operating points due to the activation of the MPC signal near the reference point. Moreover, traditional control methods generally result in slower dynamic responses, as the switching decision is based on the current measurement rather than a prediction of future states. Furthermore, in CHM-II, the average current ripple is first evaluated and then adjusted using a scaling factor that must be precisely tuned for effective operation. The second concern relates to the hysteresis-based switching method, which can result in considerable overshoots at certain operating points, particularly when the MPC signal is triggered near the reference value. The performance of this technique is highly sensitive to the selection of the hysteresis band limits—both upper and lower. Improper band settings may degrade system dynamics and lead to instability. Although approaches such as using average current ripple and ANN have been employed to enhance stability and

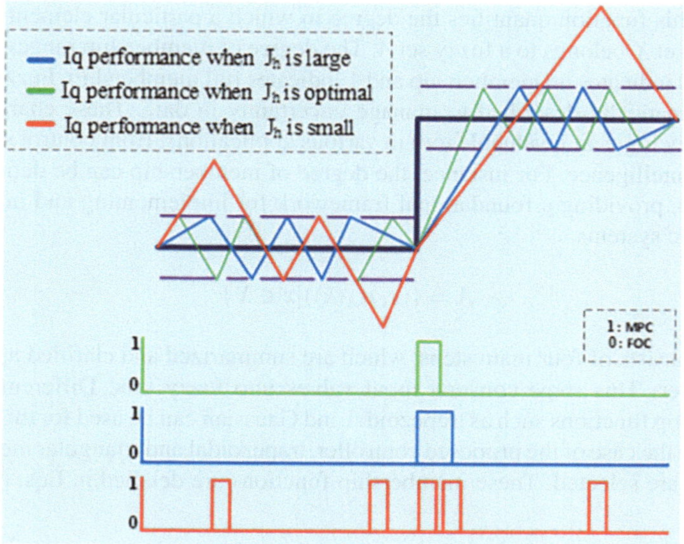

Fig. 4.29 Impact on I_q performance with different band values

improve dynamic response, they fall short in fully addressing the overshoot issue. This limitation largely stems from the difficulty in determining an optimal hysteresis band configuration that avoids triggering MPC near the reference point.

Designing an optimal hysteresis switching strategy becomes increasingly complex when multiple variables must be taken into account. This design challenge is more intricate than that of MPC, which can simultaneously manage several objectives through a unified cost function. Moreover, CHMs face limitations in two-level inverter systems, particularly when aiming to balance fast dynamic response with good steady-state behavior. In contrast, the GHM approach overcomes the need for numerous voltage vectors in multilevel inverters and eliminates the need for extensive tuning of weighting factors typically required for capacitor voltage balancing in systems with many VVs.

4.4.2 Hybrid Fuzzy-ANN-Based Current Control Methods

FL is a powerful approach to data processing based on the principles of fuzzy set theory, where truth values are represented on a range between 0 and 1. This method allows for more flexible and intuitive handling of data compared to traditional binary logic systems, where variables must be either 0 or 1. In the context of this study, FL is detailed in sources [6–8] and employed to enhance data processing and decision-making. Central to fuzzy logic is the concept of the membership function, denoted by μA. This function quantifies the degree to which a particular element X from a universal set X belongs to a fuzzy set A. The degree of membership ranges from 0 to 1, where 0 indicates no membership and 1 indicates full membership. Fuzzy sets can be mathematically described to manage uncertainty in data. These characteristics make fuzzy logic an invaluable tool in various applications, from control systems to artificial intelligence. For instance, the degree of membership can be defined using Eq. (4.21), providing a foundational framework for implementing and interpreting fuzzy logic systems.

$$A = \{(x, \mu_A(x))|x \in X\}. \tag{4.21}$$

FL1 consists of four main steps, which are summarized and clarified as follows:

Fuzzifier: This stage converts input values into fuzzy sets. Different types of membership functions such as trapezoidal and Gaussian can be used for this transformation. In the case of the proposed controller, trapezoidal and triangular membership functions are selected. These membership functions are detailed in Eqs. (4.22) and (4.23).

$$\mu_A(x) = \begin{cases} \frac{x-A}{B-A}, A < x < B \\ P,\mathrm{B} \le \mathrm{x} \le \mathrm{C} \\ \frac{D-x}{D-C}, C < x < F \\ 0, \text{otherwise} \end{cases}, \tag{4.22}$$

$$\mu_A(x) = \begin{cases} 0,\ \mathit{if}\ x \le A \\ \frac{x-A}{B-A}\ \mathit{if}\ A \le x \le B \\ \frac{C-x}{C-B}\ \mathit{if}\ B \le x \le C \\ 0\ \mathit{if}\ x \ge\ C \end{cases}. \tag{4.23}$$

Due to their computational efficiency and implementation simplicity, trapezoidal and triangular membership functions were selected for real-time system applications. Linear interpolation techniques are utilized to calculate membership degrees for input values within trapezoidal functions, as outlined in Eq. (4.24).

$$\mu_A(x) = \mu_A(X_1) + \frac{\mu_A(X_2) - \mu_A(X_1)}{X_2 - X_1}.(x - X_1). \tag{4.24}$$

Rule-base: Within fuzzy rule-based systems, two principal types of if–then rules are recognized: Mamdani and Takagi–Sugeno-Kang (TSK). The fundamental distinction lies in their approaches to deriving crisp outputs from fuzzy inputs. For this study, the Mamdani methodology has been selected [6].

Defuzzification: During defuzzification, fuzzy sets are converted to crisp output values that facilitate decision-making. The resulting crisp value enables optimal selection of switching points between MPC and FOC control modes. The centroid approach, described in Eq. (4.25), serves as the defuzzification method in this work.

$$g = \frac{\sum\limits_{i=n} x_i \mu_A(x_i)}{\sum\limits_{i=n} \mu_A(x_i)} \tag{4.25}$$

Figure 4.30 illustrates the complete block diagram of the GHM [9]. The system integrates three primary stages. Initially, fuzzy logic is applied to determine the optimal VV value using rotational speed (ω) and V_{dc} as inputs. Nevertheless, the demand for minimal computation time in switching applications presents implementation challenges for FL in real-time systems. Subsequently, the ANN undergoes training using the optimal values generated directly by FL. The FL-ANN hybrid approach delivers several key advantages: computation time reduction, improved learning flexibility and interpretability, and enhanced operational efficiency. For real-time applications, ANN utilizes simple mathematical equations with pre-determined optimal weights and biases, achieving faster computation than FL. This computational efficiency directly contributes to improved dynamic system response. Recent studies have reported hybrid switching methods with computation times as low as 1 to 10 μs [5].

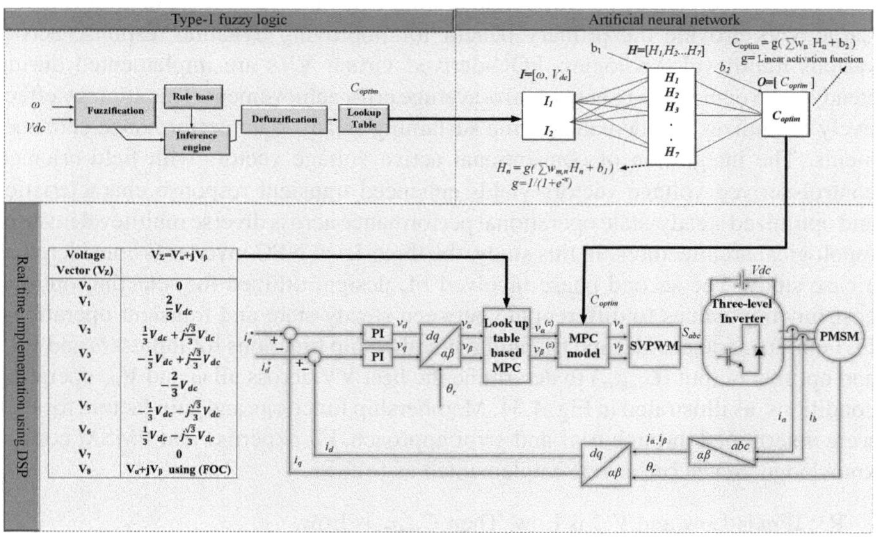

Fig. 4.30 Overall block diagram illustrating the complete process of the generalized hybrid method (GHM) using FL and ANN

The FL-ANN improves learning adaptability and system interpretability, resulting in enhanced overall performance. While FL serves as an effective decision-making framework capable of managing uncertainty for optimal VV selection, it lacks the learning capabilities of ANN, which efficiently acquires input–output relationships Moreover, ANN can learn and model the non-linear relationship between input and output. Such functionality supports generalization to novel input data and allows for diverse system implementations using transfer learning approaches.

While high accuracy can be achieved with advanced machine learning methods, their high computation time may impact dynamic performance. Thus, to tackle this issue, an low structure of Multi-layer neural network (MLNN) is used. The MLNN stands out by delivering speed and accuracy, which makes it particularly advantageous for real-world applications. Additionally, it is well-suited for use with low-order controllers like general DSPs. This adaptability ensures that the MLNN can be used in a wide range of scenarios, providing significant versatility and flexibility.

The MPC cost function was directed by optimal values from the ANN to facilitate optimal VV selection. The preliminary design phase encompassed the integration of FOC architecture, wherein the stationary reference frame voltage components ($V\alpha$, $V\beta$) generated through FOC transformation algorithms served as supplementary voltage vectors, as illustrated in Fig. 4.30. The proposed lookup table-based model predictive control methodology employs a finite set of nine voltage space vectors comprising: two zero-voltage vectors (V0, V7), six active voltage vectors (V1-V6), and one virtual voltage vector synthesized from FOC outputs. This tabular MPC implementation avoids the computational complexity associated with large number of voltage vector synthesis while maintaining equivalent steady-state performance. Large VVs provide the primary benefit for improving dynamic response across various multilevel topologies. FOC-derived virtual VVs are implemented during steady-state operations to ensure zero-average error achievement. This strategy effectively minimizes VV demands while sustaining steady-state performance enhancements. The integration of conventional active voltage vectors with field-oriented control-derived voltage vectors yields enhanced transient response characteristics and optimized steady-state operational performance across diverse multilevel inverter topological architectures. In this study, the three-level NPC inverter is considered as a case study. The second phase involved FL design, utilized for selecting optimal cost function values to differentiate between steady-state and transient operations. FL1 employs trapezoidal and triangular membership functions for inputs (ω and V_{dc}) and optimal output (C_{optim}) to determine the best VV across all ω and V_{dc} operating conditions, as illustrated in Fig. 4.31. Membership functions and gain factors for FL1 were determined through trial-and-error approach, FL expertise and PMSM control knowledge. Seven rules were implemented as follows.

R^1: IFω is Low and V_{dc} is Low, Then C_{optim} is Low.
R^2: IFω is Medium and V_{dc} is Medium, Then C_{optim} is Medium.
R^3: IF ω is High and V_{dc} is High, Then C_{optim} is High.
R^4: IF ω is Low and V_{dc} is Medium, Then C_{optim} is Medium.
R^5: IF ω is Medium and V_{dc} is Low, Then C_{optim} is Low.

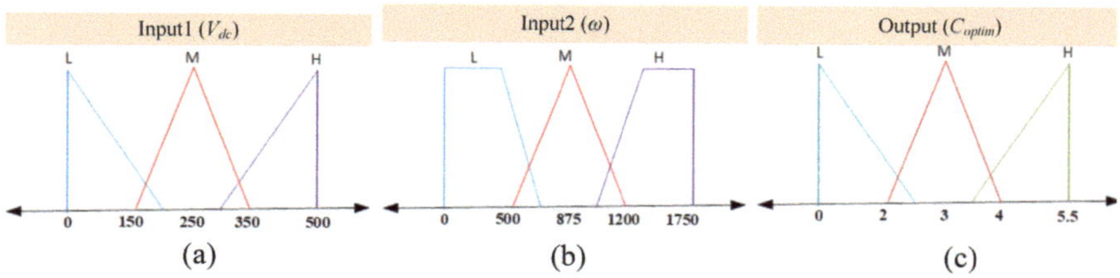

Fig. 4.31 Membership function: **a** Vdc-link voltage (Input1); **b** speed (Input2) (ω_c); **c** optimal value C_{optim} (output)

R^6: IF ω is Medium and V_{dc} is High, Then C_{optim} is High.
R^7: IF ω is High and V_{dc} is Medium, Then C_{optim} is Medium.

According to the FL rules, V_{dc} provides greater influence than ω for C_{optim} selection. Figure 4.32 displays the mathematical and design framework of the MLNN [4]. The weights in the MLNN are updated using the Levenberg–Marquardt algorithm. The data is split into 70% for training and 30% for testing the performance of the MLNN model. With a single hidden layer, MLNN aims for efficient execution. The approach delivers excellent overall performance, reaching a correlation coefficient of roughly 98%. This degree of precision is sufficient for the optimal value to approximate the exact value adequately. When designing an ANN to operate with various motors of differing rated power, training data must encompass both motor types for effective differentiation. Furthermore, enhancing the robustness of an ANN across different motors entails incorporating additional variables such as motor type, rating, and efficiency as inputs. These augmentations help distinguish between the distinct characteristics of various motors, ensuring the adaptability and efficacy of the ANN. The optimal VV was selected by MPC using the optimal value, as illustrated in Eq. (4.26).

$$\begin{cases} V_{1-7}, \ CF > C_{optim} \\ V_8, \ CF < C_{optim} \end{cases}, \tag{4.26}$$

In scenarios where CF exceeds C_{optim}, which indicates that the predicted system is in a transient state, the controller employs the VVs generated by MPC for rapid tracking of its reference. Conversely, when CF is less than C_{optim}, indicating the system is in a steady state, the controller selects the VV generated by FOC. It is important to highlight that the selection of optimal VVs is based on the same cost function in both cases, ensuring that no additional computation time is required. This approach streamlines decision-making and maintains efficiency in control operations. Figure 4.33 illustrates the differences between the control diagrams of the GHM method and CHMs. The GHM utilizes the system current model to predict the next current value, resulting in a fast dynamic response and minimal overshoot. This technique integrates the MPC cost function, which relies on predicted current values, with the optimal values from FL1-ANN to achieve improved VV selection. In

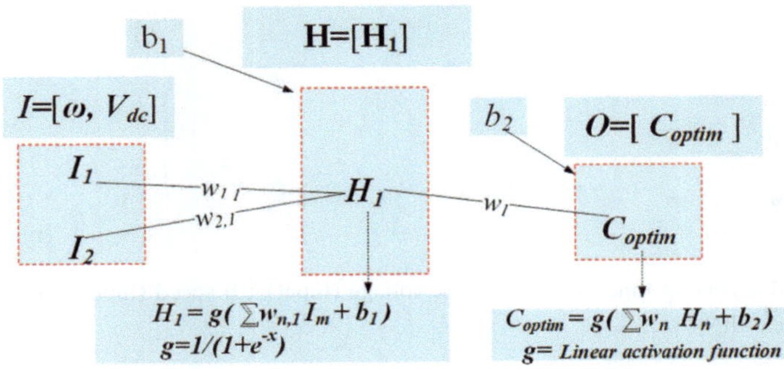

Fig. 4.32 Structure of the MLNN implemented in the GHM

contrast, CHMs use only instantaneous current values and hysteresis-based switching between FOC and MPC, leading to degraded dynamic performance and increased overshoot.

A major limitation of CHMs lies in their hysteresis switching methodology, which generates substantial overshoots at specific operating points when MPC activation occurs near reference operating conditions. Hysteresis switching performance depends critically on hysteresis band values. Wide "high bands" produce slow dynamics, whereas narrow "low bands" create unstable behavior.

Furthermore, optimal hysteresis switching becomes increasingly difficult when multiple variables are involved, complicating the design methodology. This contrasts with MPC cost functions, which integrate multiple objectives more effectively. CHMs are also restricted to two-level inverter system for achieving rapid dynamic response with satisfactory steady-state performance. In contrast, the GHM removes the requirement for extensive VVs in multilevel inverters, improving steady-state performance while eliminating complex weighting factor adjustments needed for capacitor balancing when managing large number of VVs.

4.4.3 Experimental Validation

The setup outlined in Fig. 4.34 was employed to experimentally test the effectiveness of the GHM. In this configuration, a 5 kW PMSM was mechanically linked to an IM. The system utilized a TMS320F28335 from Texas Instruments. Detailed experimental parameters can be found in Table 4.1.

Different cases were used to evaluate the performance of GHM, CHMs, and FOC. They are summarized as follows.

(1) Case A: $\omega = 300$ r/min, $V_{dc} = 100$ V, and I^*_q change from 1 to 4 A.
(2) Case B: (step up test) $\omega = 450$ r/min, $V_{dc} = 150$ V, and I^*_q change from 1 to 5 A.

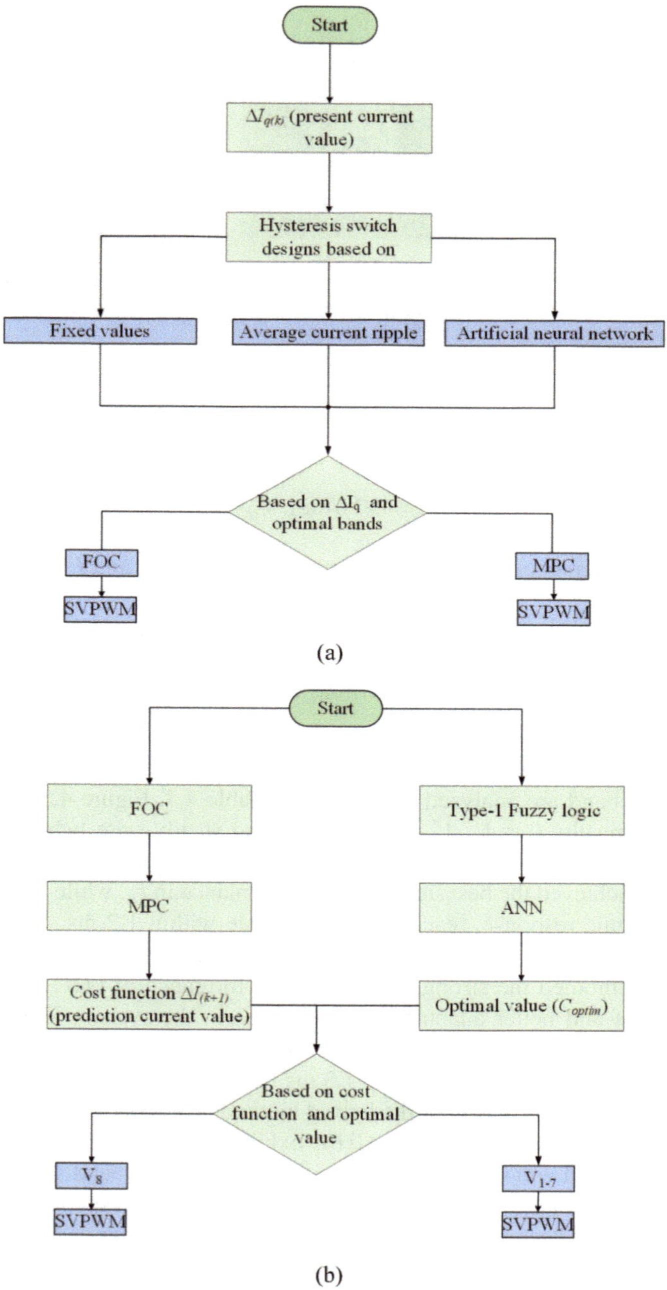

Fig. 4.33 Flowcharts of hybrid methods: **a** CHM; **b** GHM

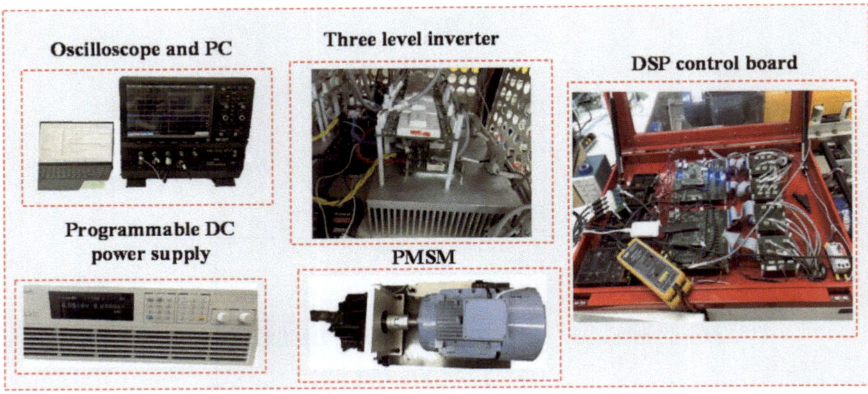

Fig. 4.34 Experimental setup

(3) Case C: (step down test): $\omega = 450$ r/min, $V_{dc} = 150$ V, and I^*_q change from 5 to 1 A.
(4) Case D: (overshoot test) $\omega = 150$ r/min, $V_{dc} = 200$ V, and I^*_q change from 1 to 7 A.
(5) Case E: (large dynamic) $\omega = 900$ r/min, $V_{dc} = 450$ V, and I^*_q change from 1 to 15 A.
(6) Case F: (High speed) $\omega = 1500$ r/min, $V_{dc} = 500$ V, and I^*_q change from 7 to 14 A.

Furthermore, to assess the robustness of the hybrid method, three conditions of the switching band are analyzed as detailed in Table 4.7. Figure 4.35 presents the experimental results for: I_q, I^*_q, SS, and I_A, and steady-state behavior for Case A, comparing five control strategies: FOC, MPC, CHMI, CHMII, and GHM. The FOC method achieved the best steady-state performance in I_A, while MPC delivered superior dynamic response, reaching the reference within 1.2 ms. In comparison, FOC's dynamic response was slower at 3.1 ms. The CHMI, CHMII, and GHM approaches combined the strong steady-state performance of FOC with enhanced responsiveness and better adaptability to changes in the system. However, GHM showed faster dynamics similar to those of MPC, with a response time of 1.2 ms, compared to response times of 1.5 and 1.3 ms for CHMI and CHMII, respectively.

A performance comparison of the hybrid methods was conducted to clarify their behavior using the switching band. GHM exhibited faster dynamic performance compared to CHMI and CHMII due to its smaller switching band size, which typically indicates quicker dynamics. GHM also adapted and responded earlier than CHMI and CHMII because it utilizes the predictive capabilities of MPC, unlike the CHMs which rely on hysteresis switching. Cases B and C evaluated the step-up and step-down scenarios of the I_q current. The FOC method demonstrated slower dynamic performance, especially when I_q is increases from 1 to 5 A. In contrast, the GHM method showed significantly faster performance compared to other methods,

Table 4.7 Impact of the switching band on the system dynamic performance

Evaluation types of Switching band	Type	Dynamic	Overshoot
Size	Small	Fast	–
	Medium	Medium	
	Large	Slow	
Prediction	K + 1	Fast	–
	K	Medium	
	K–1	Slow	
Reference-end	Before reference	–	Low
	At reference		Medium
	After reference		Large

achieving a response time of 1.1 ms, whereas FOC, CHMI, and CHMII had response times of 2.9, 2, and 1.8 ms, as can be seen in Fig. 4.36 shows this fast performance. This happens because the GHM switching band is smaller, which starts working before the other CHM bands. Basically, GHM can start adjusting and responding to system changes earlier than the other methods. Therefore, the GHM method has the fastest dynamic performance. In a step-down test, CHMI and CHMII had larger overshoots of 1.6 A and 1 A, as shown in Fig. 4.37b, c. FOC had no overshoot, but it also had the slowest dynamic performance at 3.5 ms. In contrast, Fig. 4.37d highlights that GHM produced minimal overshoot 0.5 A together with the fastest transient response 0.3 ms. This advantage stems from the predictive strength incorporated through MPC. Another observation is that the switching band of GHM completed earlier than the target reference, leading to reduced overshoot compared with CHMs. Case D, summarized in Fig. 4.38, was intended to emphasize the limitations of CHMs. Both CHMI and CHMII shown overshoots larger than those of FOC and GHM, with peak values of 2.5 A and 1.9 A, respectively. On the other hand, FOC and GHM maintained comparatively small overshoot levels, although FOC required a longer settling time of 2.2 ms, whereas GHM responded more quickly at 0.9 ms. To assess performance under demanding transient conditions, Case E was performed as displayed in Fig. 4.39. Here, GHM again showed superior dynamic speed, responding in 1.2 ms, while FOC and CHMII followed at 2 and 1.4 ms. CHMI could not be subjected to this test due to its excessive overshoot, which posed safety risks. In this scenario, FOC and GHM limited overshoot to 1 A, while CHMII reached 2 A. Further benchmarking was carried out in a high-speed, high-current operating point, with motor speed fixed at 1500 rpm and Iq stepped from 7 to 14 A, as given in Fig. 4.40. Results confirmed that GHM consistently outpaced the alternatives, achieving a 0.9 ms transient, compared with 1.8 ms for FOC and 1.3 ms for CHMII. Summarizing across all cases, GHM demonstrated negligible steady-state error, rapid dynamic behavior, and minimal overshoot. The method was therefore validated against well-established industrial strategies such as FOC, MPC, and the recent CHMs, confirming both its efficiency and its practicality.

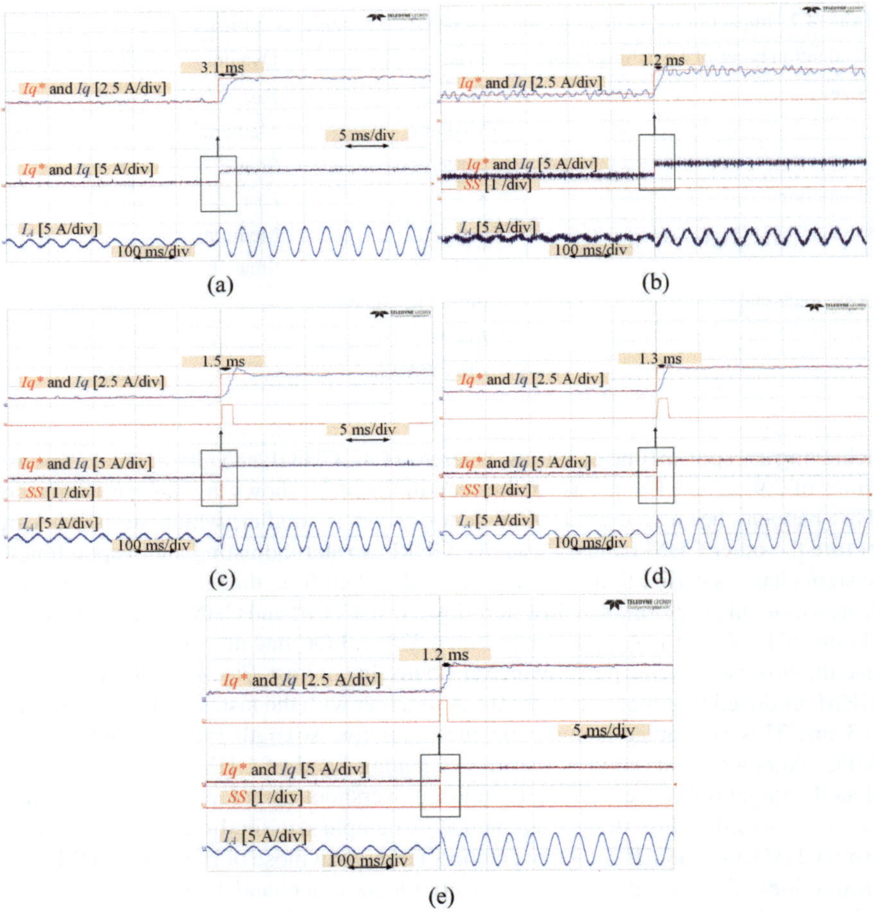

Fig. 4.35 Experimental evaluation of GHM with various methods at 300 r/min for I^*_q change from 1 to 4 A: **a** field oriented control (FOC); **b** model predictive control (MPC); **c** based on hysteresis CHMI; **d** based on average current ripples (CHMII); **e** generalized hybrid method (GHM)

Table 4.8 presents the computation times for the conventional and GHM methods. It is apparent that MPC used a higher computation time compared to FOC. This is due to the optimization of the cost function required to find the optimal VV. In CHMI and CHMII, the total delay was influenced by both the FOC/MPC stage and the additional switching calculation. The recorded maximums were 18 μs for CHMI and 19 μs for CHMII. The slightly higher value of CHMII comes from the inclusion of average ripple estimation. By contrast, GHM produced a fixed runtime equal to the combined costs of FOC, MPC, and ANN, giving 28 μs overall. Although longer, this duration remains practical for real-time execution.

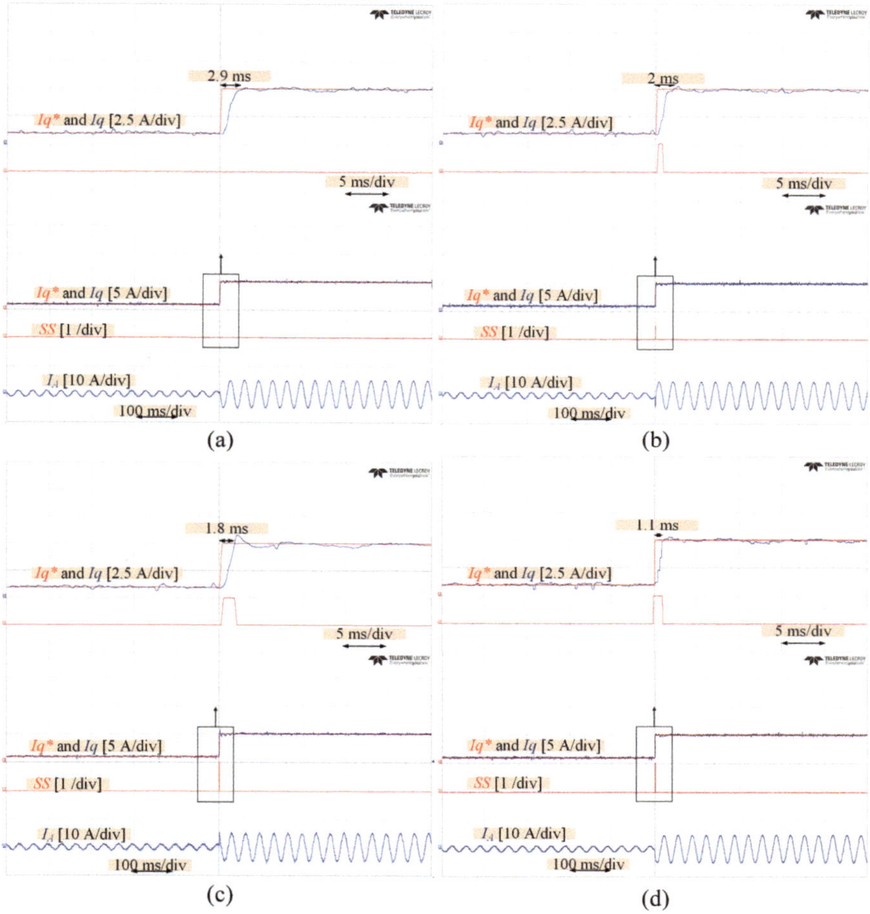

Fig. 4.36 Experimental evaluations of the GHM with various methods at 450 r/min with I^*_q change from 1 to 5 A: **a** field oriented control (FOC); **b** based on hysteresis CHMI; **c** based on average current ripples (CHMII); **d** generalized hybrid method (GHM)

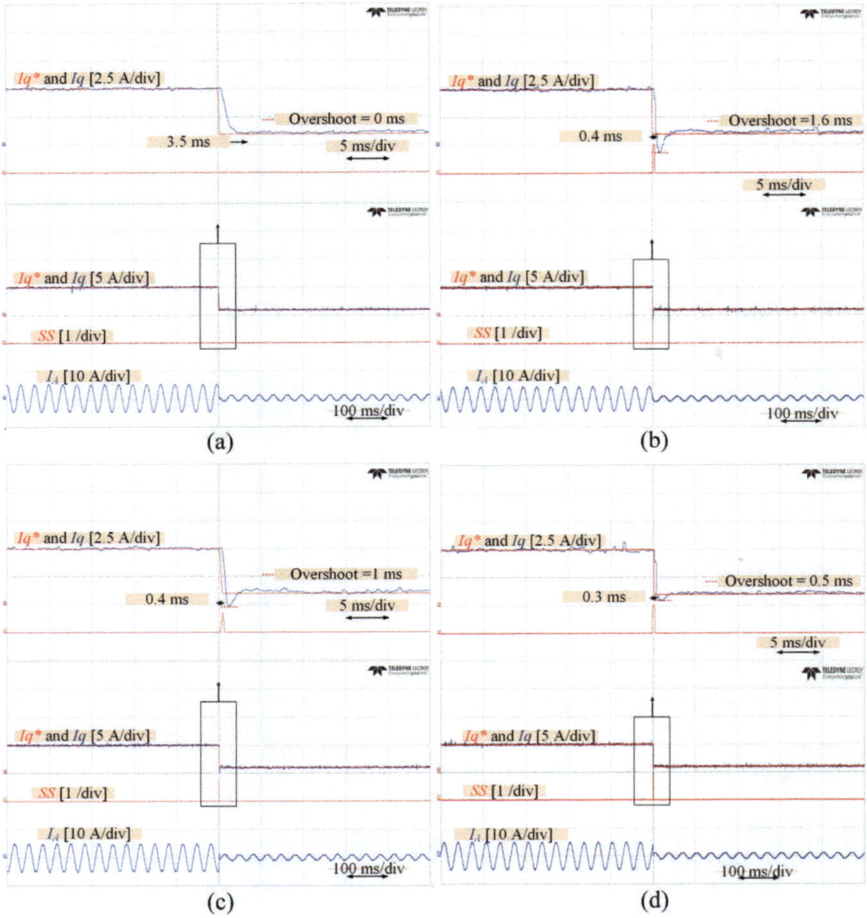

Fig. 4.37 Experimental evaluations of the GHM with various methods at 450 r/min with I^*_q change from 5 to 1 A: **a** field-oriented control (FOC); **b** based on hysteresis CHMI; **c** based on average current ripples (CHMII); **d** generalized hybrid method (GHM)

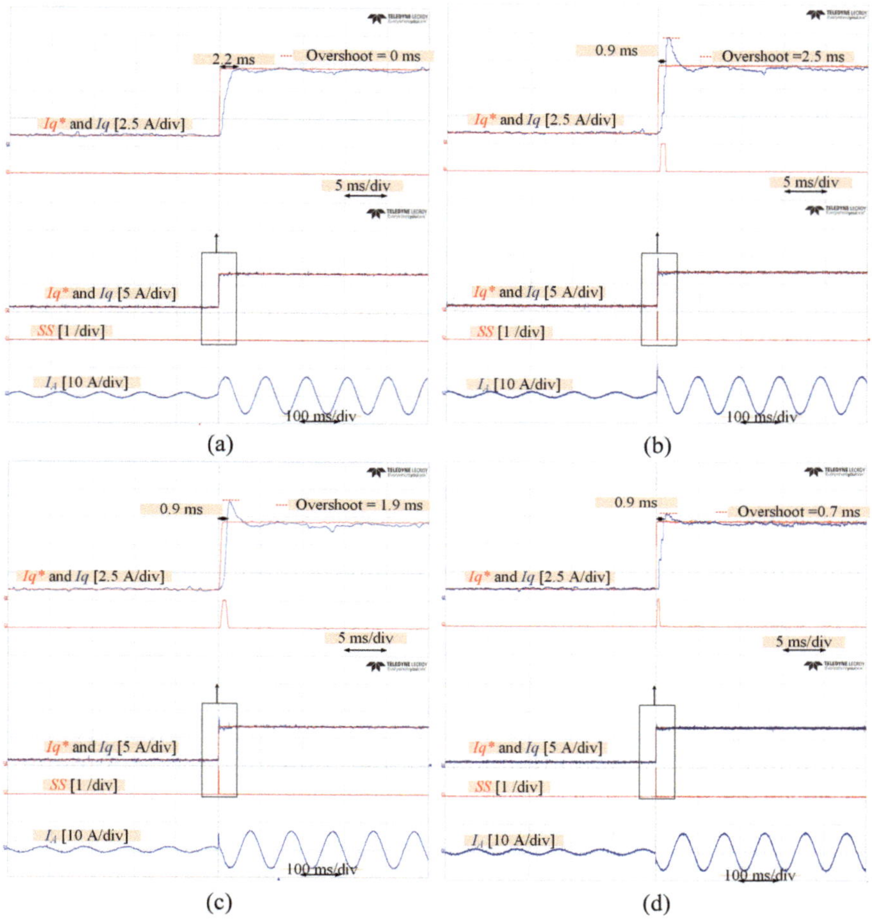

Fig. 4.38 Experimental evaluations of the GHM with various methods at 150 r/min with I^*_q change from 1 to 7 A: **a** field-oriented control (FOC); **b** based on hysteresis CHMI; **c** based on average current ripples (CHMII); **d** generalized hybrid method (GHM)

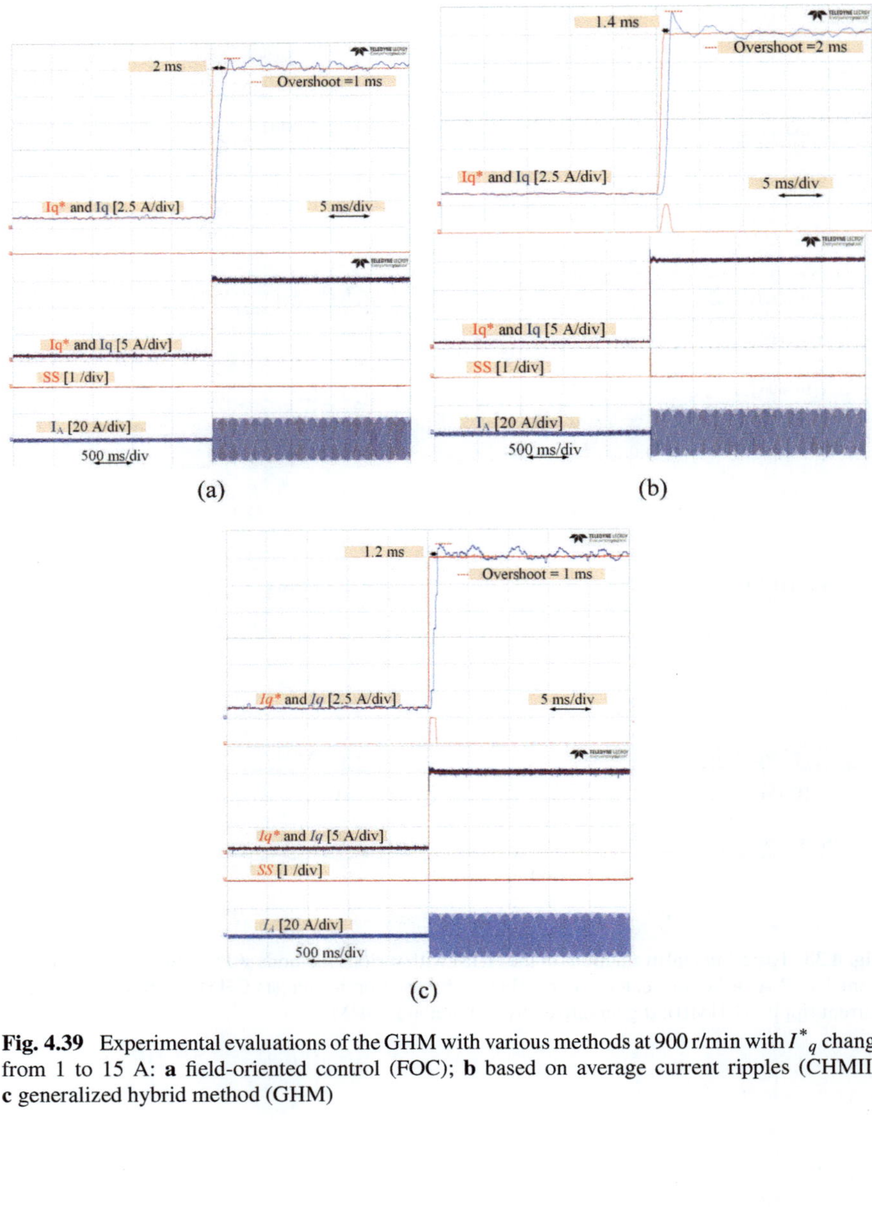

Fig. 4.39 Experimental evaluations of the GHM with various methods at 900 r/min with I^*_q change from 1 to 15 A: **a** field-oriented control (FOC); **b** based on average current ripples (CHMII); **c** generalized hybrid method (GHM)

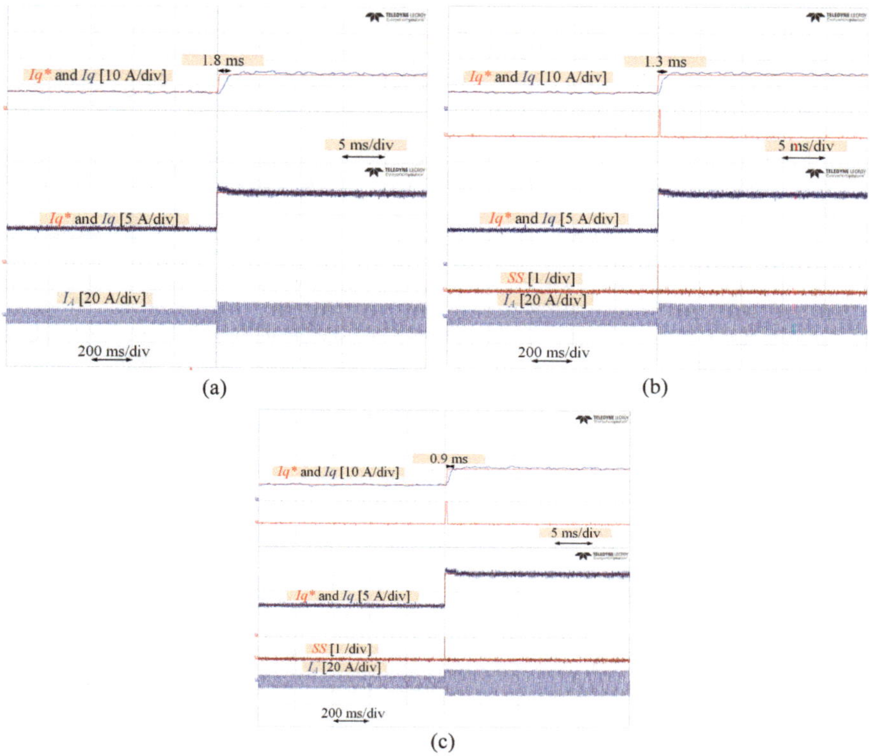

Fig. 4.40 Experimental evaluations of the GHM with various methods at 1500 r/min with I^*_q change from 7 to 14 A: **a** field-oriented control (FOC); **b** based on average current ripples (CHMII); **c** generalized hybrid method (GHM)

Table 4.8 Comparison of execution times

Method	Computation time (μs)
FOC	4
MPC	17
CHMI	18
CHMII	19
GHM	28

References

1. Kim S-H (2017) Electric motor control: DC, AC, and BLDC motors. Elsevier, Amsterdam, Netherlands
2. Al-kaf HA, Chia KS, Alduais NA (2018) A comparison between single-layer and multilayer artificial neural networks in predicting diesel fuel properties using near-infrared spectrum. Pet Sci Technol 36(6):411–418

3. Al-Kaf HAG, Lee K-B (2022) Low complexity MPC-DSVPWM for current control of PMSM using neural network approach. IEEE Access 10:132596–132607
4. Wang D, Shen ZJ, Yin X, Tang S, Liu X, Zhang C, Wang J, Rodriguez J, Norambuena M (2022) Model predictive control using artificial neural network for power converters. IEEE Trans Ind Electron 69(4):3689–3699
5. Al-Kaf HAG, Hakami SS, Lee K-B (2022) Hybrid current controller for permanent-magnet synchronous motors using robust switching techniques. IEEE Trans Power Electron 38(3):3711–3724
6. Wang S-C, Liu Y-H (2011) A modified PI-like fuzzy logic controller for switched reluctance motor drives. IEEE Trans Ind Electron 58(5):1812–1825
7. Chen J, Xu C, Wu C, Xu W (2016) Adaptive fuzzy logic control of fuel-cell-battery hybrid systems for electric vehicles. IEEE Trans Ind Inf 14(1):292–300
8. Soliman MA, Hasanien HM, Azazi HZ, El-Kholy EE, Mahmoud SA (2019) An adaptive fuzzy logic control strategy for performance enhancement of a grid-connected PMSG-based wind turbine. IEEE Trans Ind Inf 15(6):3163–3173
9. Al-Kaf HAG, Halabi LM, Lee K-B (2024) A generalized integrated MPC-fuzzy-neural network approach for multilevel inverter-fed PMSMs. IEEE Trans Ind Inf 20(2):2751–2761

Chapter 5
Machine Learning Methods for Fault Detection

Abstract This chapter explores machine learning based fault detection strategies for three-level Neutral Point Clamped (NPC) inverters, which are widely used in power electronics due to their reduced harmonic distortion and lower power losses. Despite these advantages, NPC inverters remain vulnerable to faults—particularly open-circuit faults, which are more challenging to detect than short-circuit faults. To address this, this chapter focuses on leveraging ML algorithms to classify and detect faults using historical and real-time system data. Key ML techniques discussed include artificial neural networks (ANNs), and ensemble machine learning (EML) methods. EML and ANNs are highlighted for their robustness and accuracy in identifying various fault types. The chapter also emphasizes the role of explainable ML to ensure transparency and trust in fault detection systems for industrial applications. Furthermore, the implementation of ML-based detection on digital signal processors (DSPs) is examined, demonstrating the practical feasibility of real-time applications. Overall, the chapter highlights how ML enhances fault resilience and operational reliability in power electronic systems.

5.1 NPC Inverter Modeling

The circuit diagram in Fig. 5.1 illustrates a standard three-phase three-level NPC inverter with twelve Si IGBT switching devices and six clamping diodes. The inverter features a dual-capacitor dc-link (C1 and C2) separated by a neutral point O. Clamping diodes connect each phase output to O, enabling its application in medium-voltage and high-voltage systems. Notably, the configuration employs four Si IGBTs in series, resulting in 27 switching states for robust control as well as reduced ripples and losses compared to conventional six-switch inverters. The output voltage (V_{xO}) is three-level according to expression (5.1).

H. A. G. Al-kaf and K.-B. Lee, *Machine Learning Integration in Power Electronics*,
https://doi.org/10.1007/978-981-95-3845-4_5

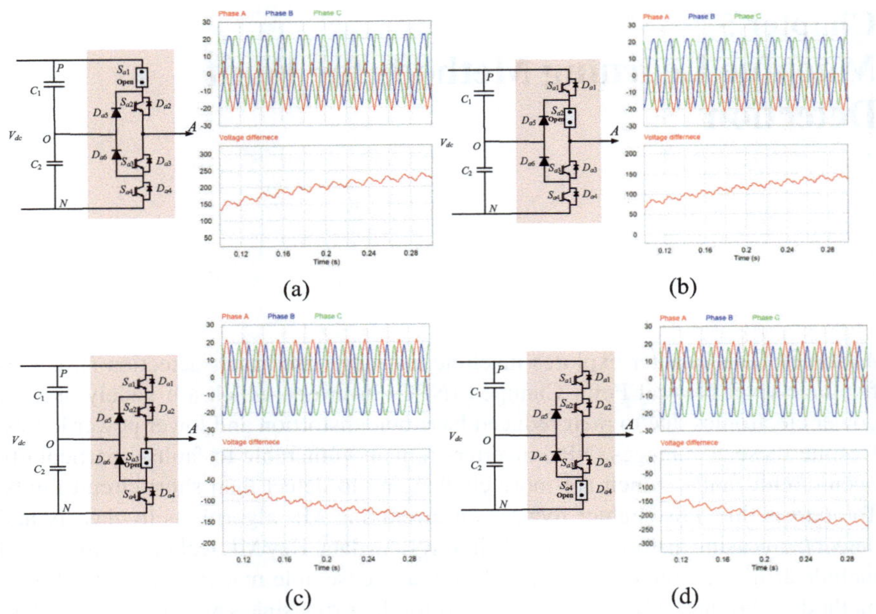

Fig. 5.1 Three-phase currents waveforms and voltage differences for phase A under various fault types: **a** S1; **b** S2; **c** S3; **d** S4

$$\left\{\begin{array}{l} S = (S_a, S_b, S_c),\ S_{a,b,c} \in \{-1, 0, +1\} \\ P_{state}(+1) \rightarrow S_{x1}, S_{x2} \text{ are ON} \rightarrow V_{xO}= +V_{dc}/2 \\ O_{state}(0) \rightarrow S_{x2}, S_{x3} \text{ are ON} \rightarrow V_{xO}=0 \\ N_{state}(-1) \rightarrow S_{x3}, S_{x4} \text{ are ON} \rightarrow V_{xO}= -V_{dc}/2 \end{array}\right\}, \tag{5.1}$$

where x represents the three phases a, b, and c.

Figure 5.1 illustrates the current flow patterns and operational states of the three-level NPC inverter, namely the P, O, and N states. In the P state, S1 and S2 are activated, allowing current to flow either from the upper arm to the load or from the load through the diode to the positive terminal. States O and N exhibit distinct activation configurations for these switches, leading to current flow variations. In the event of an open-circuit fault, the output voltage (V_{xO}) of the three-level NPC inverter can experience distortion due to the absence of voltage states, depending on the fault location [1]. For instance, if an open-switch fault occurs in S1, the system operates in the state P, where the faulty phase current has a smaller amplitude compared with the other two phases. In the second scenario, with an open-switch fault in S2, the corresponding faulty positive-phase current is nearly zero, necessitating consideration of the operation states P and O. A third case arises when an open-circuit fault occurs in S3, resulting in an approximately zero faulty negative-phase current.

Consequently, the operational states O and N should be taken into account. The final case involves an open-switch fault in S4, leading to the state N operation, where the faulty phase current has a smaller amplitude compared to the other two phases.

5.2 Ensemble Machine Learning Methods

5.2.1 Types of Ensemble Machine Learning (EML) Methods

EML combines multiple models to leverage their strengths and overcome individual weaknesses, resulting in a more robust and accurate model. EML techniques such as boosting, bagging, and stacking are widely used. Bagging, including methods such as bootstrap aggregating, involves training multiple models on randomly sampled subsets of the data and combining their predictions through voting or averaging, thereby reducing overfitting and enhancing accuracy. In contrast, boosting methods, such as AdaBoost, focuses on sequentially training models, with each iteration correcting the mistakes of the previous one. Stacking trains a meta-model to combine predictions from base models for better results.

Various EML methods have been applied for fault detection in NPC inverters. The widely recognized Random Forest (RF) which combines multiple decision trees, each trained on random data subsets, to improve model robustness. AdaBoost enhances weaker models by assigning more weight to previously misclassified instances, while gradient boosting models such as GBM and XGBoost use gradient descent to optimize models over multiple iterations. Each technique brings distinct advantages, such as the resistance to overfitting in RF and the ability to focus on difficult data points in AdaBoost [2].

Other methods, such as LightGBM and XGBoost provide optimized and adaptable solutions for processing large datasets. LightGBM employs a leaf-wise growth strategy for balanced tree structures, while XGBoost optimizes gradient boosting through regularization and novel tree construction techniques. Additionally, a voting classifier, which combines multiple algorithms such as RF, Gaussian Naive Bayes, and AdaBoost, enhances prediction accuracy by averaging the predictions of these individual classifiers using a soft voting method. These EML techniques collectively improve the fault detection process in power electronics.

Current and voltage data for EML methods were acquired through both simulations and physical experiments, covering current levels from 5 to 15 A and voltage ranges between 100 and 400 V. Measurements were taken across several cycles during normal operation and under single switch fault conditions among thirteen switches. This procedure was repeated for each current level across various fault types. Simulated data were produced using PSIM software, while experimental data were extracted from oscilloscope readings saved in CSV format. The dataset was divided into 80% for training and 20% for testing. For both datasets, the three-phase currents and the difference in DC-link voltage were normalized using the min–max method.

Various evaluation metrics are employed to assess the performance of the EML methods such as recall, accuracy F1score and precision. The definitions of these metrics are as follows:

$$\text{Accuracy(Acc)} = \frac{TP + TN}{TP + TN + FP + FN} \tag{5.2}$$

$$\text{Precision(Pr .)} = \frac{TP}{TP + FP} \tag{5.3}$$

$$\text{Recall(Re.)} = \frac{TP}{TP + FN}, \tag{5.4}$$

$$F1\text{ - score} = \frac{2 * \text{Re.} * \text{Pr .}}{\text{Re.} * \text{Pr .}}, \tag{5.5}$$

where *TP* represents true positive (correctly identified faults), *FP* represents false positive (incorrectly identified faults), *TN* represents true negative (correctly identified non-faults), and *FN* represents false negative (missed faults). All results reported correspond to the performance on the test dataset.

5.2.2 Simulation Validation

Data for the EML methods was generated using PSIM, with simulation parameters detailed in Table 5.1. The methods were implemented using Python libraries such as scikit-learn, LightGBM, and XGBoost. The network was trained offline on a preprocessed dataset, with parameter tuning performed to achieve optimal performance. Figure 5.1 presents the simulated output currents corresponding to various faulty switches. Phases A, B, and C are represented by red, green, and blue waveforms, respectively. The results confirm that the three-phase currents preserve their periodic nature under different fault types. Power converter faults create distinct distortions in output waveforms, with three-phase load currents experiencing the most pronounced effects. These faults additionally cause voltage imbalances across the upper and lower DC-link capacitors. Furthermore, fault conditions result in substantial increases in total harmonic distortion (THD) levels.

Table 5.1 System parameters

System parameters	Value
V_{DC}	600 [V]
R	10 [Ω]
L	2 [mH]
f_{fund}	50 [Hz]
f_{sf}	30 [kHz]

Notably, certain fault conditions can produce nearly identical current waveform patterns. Consequently, acquiring current waveform data from a single period is insufficient to ensure reliable fault diagnosis, especially when the sampling strategy captures only one data point per sampling interval. To address this limitation, supplementary inputs including DC-link voltage differences and time features were introduced. The performance of the EML methods with different feature sets is presented in Table 5.2. The results show that relying solely on three-phase currents leads to lower classification performance, with the highest accuracy reaching 0.63 using RF, AdaBoost, and Bagging. In contrast, using only time features and the DC-link voltage difference results in improved performance, with bagging achieving the highest accuracy of 0.92. The ability to distinguish fault types stems from analyzing the magnitude and polarity of the DC-link voltage difference. Additionally, examining the timing and shape of the phase current waveforms helps detect open-circuit faults in NPC inverters. Building on these results, the proposed approach utilizes a combination of three-phase current signals, DC-link voltage difference, and time-based features as inputs for the EML methods.

All methods, except Gradient Tree Boosting, demonstrate strong accuracy above 0.94. Among them, the Bagging technique achieves the highest accuracy of 0.99. Table 5.3 provides the precision scores for different EML methods used to detect open-circuit faults in an NPC inverter. Precision, which measures the proportion of true positive results relative to all positive predictions, is used to evaluate performance.

The majority of algorithms demonstrate robust performance across different fault categories, achieving precision values between 0.90 and 1.00. Random Forest, Bagging, AdaBoost, and an ensemble combining three techniques consistently achieve excellent precision across various fault classifications. LightGBM and XGBoost also show strong results, albeit marginally lower than the leading approaches. Gradient Tree Boosting exhibits reduced precision for specific fault types. While all algorithms attain perfect precision (1.00) for normal operational state detection, none achieve the same precision for fault categorization. Table 5.4 presents

Table 5.2 Simulation results for the accuracy of EML methods using various input features

Methods	Time, voltage difference and three phase currents	Three phase currents	Time and voltage difference
Soft (GaussianNP + RF + Adaboost)	0.95	0.63	0.89
Bagging	0.99	0.63	0.92
AdaBoost	0.97	0.63	0.88
Gradient tree boosting	0.78	0.40	0.44
Lightgpm	0.95	0.57	0.62
Xgboost	0.94	0.59	0.58
RF	0.95	0.63	0.89

Table 5.3 Simulation results for the precision of EML methods across different open-circuit faults

Fault type	EML methods						
	Bagging	RF	AdaBoost	Gradient tree boosting	Lightgpm	Xgboost	GaussianNP + RF + Adaboost
Normal	1.00	1.00	1.00	1.00	1.00	1.00	1.00
A-phase S_1	0.98	0.96	0.96	0.68	0.94	0.93	0.96
A-phase S_2	0.99	0.92	0.97	0.75	0.94	0.94	0.93
A-phase S_3	0.99	0.92	0.98	0.82	0.96	0.93	0.93
A-phase S_4	0.98	0.95	0.96	0.66	0.93	0.94	0.96
B-phase S_1	0.99	0.94	0.96	0.66	0.91	0.92	0.93
B-phase S_2	0.98	0.89	0.96	0.75	0.92	0.91	0.88
B-phase S_3	0.98	0.94	0.97	0.82	0.95	0.94	0.93
B-phase S_4	0.98	0.96	0.96	0.67	0.93	0.94	0.96
C-phase S_1	0.98	0.96	0.96	0.66	0.93	0.93	0.95
C-phase S_2	0.99	0.94	0.98	0.77	0.95	0.94	0.92
C-phase S_3	0.98	0.92	0.97	0.79	0.92	0.90	0.90
C-phase S_4	0.99	0.95	0.96	0.66	0.91	0.93	0.94

the precision, recall, and F1-score metrics for the Bagging approach, which surpasses all competing methods. The precision values range from 0.98 to 1.00, demonstrating the method's superior capability in detecting open-circuit fault conditions.

The model achieves the highest precision in classifying the normal operating condition, attaining a perfect score of 1.00. Recall values, which indicate the model's

Table 5.4 Simulation results for the performance of the bagging method across different open-circuit faults

Type of fault	Precision	Recall	F1-score
Normal	1.00	1.00	1.00
A-phase S_1	0.98	0.99	0.99
A-phase S_2	0.99	0.99	0.99
A-phase S_3	0.99	0.99	0.99
A-phase S_4	0.98	0.98	0.98
B-phase S_1	0.99	0.98	0.98
B-phase S_2	0.98	0.98	0.98
B-phase S_3	0.98	0.98	0.98
B-phase S_4	0.98	0.98	0.98
C-phase S_1	0.98	0.98	0.98
C-phase S_2	0.99	0.99	0.99
C-phase S_3	0.98	0.98	0.98
C-phase S_4	0.99	0.98	0.98

effectiveness in detecting actual fault cases, remain consistently high across all fault types, ranging from 0.98 to 0.99. Similarly, the F1-scores, which measure the harmonic mean of precision and recall, consistently achieve high values, ranging from 0.98 to 0.99, demonstrating the strong reliability of the classification system.

5.2.3 Experimental Validation

The effectiveness of the fault-tolerant method is validated through experimental results obtained using a prototype NPC inverter. Figure 5.2 presents the experimental setup employed in this study. The configuration includes UF3SC120009 SiC Cascode MOSFET components and a TMS320F28377 digital signal processor control unit from Texas Instruments (TI). The main test conditions align with the specifications provided in Table 5.1. Figure 5.3 demonstrates open-circuit fault conditions in the NPC inverter, with emphasis on the A-phase. The faults are deliberately induced in switching components Sa1, Sa2, Sa3, and Sa4.

From Fig. 5.3, it can be seen that the behavior is identical to the simulation results for the three phase currents.

The faults produce substantial distortions in the output waveforms, particularly affecting three-phase load currents. These fault conditions also cause voltage imbalances between the upper and lower DC-link capacitors. Additionally, the faults result in considerable deterioration of THD performance. Moreover, both the three-phase

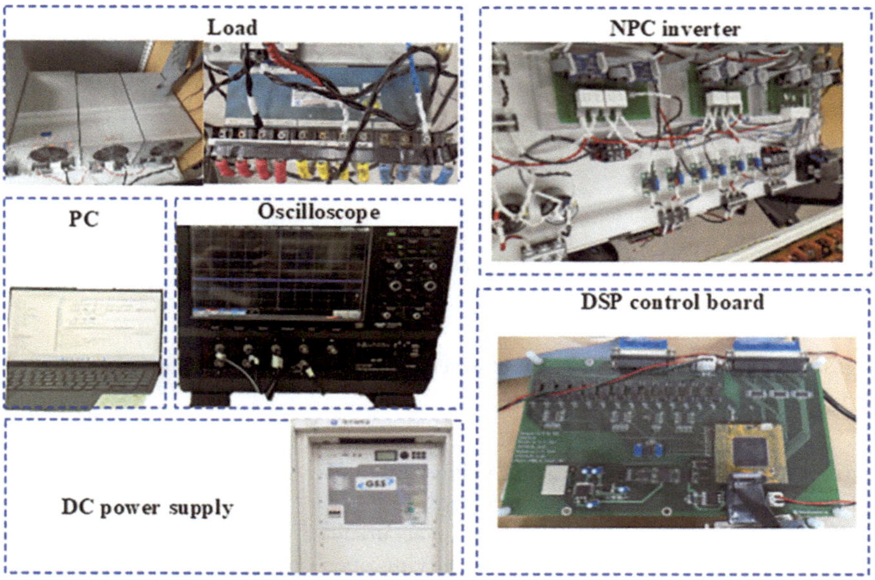

Fig. 5.2 Experimental setup

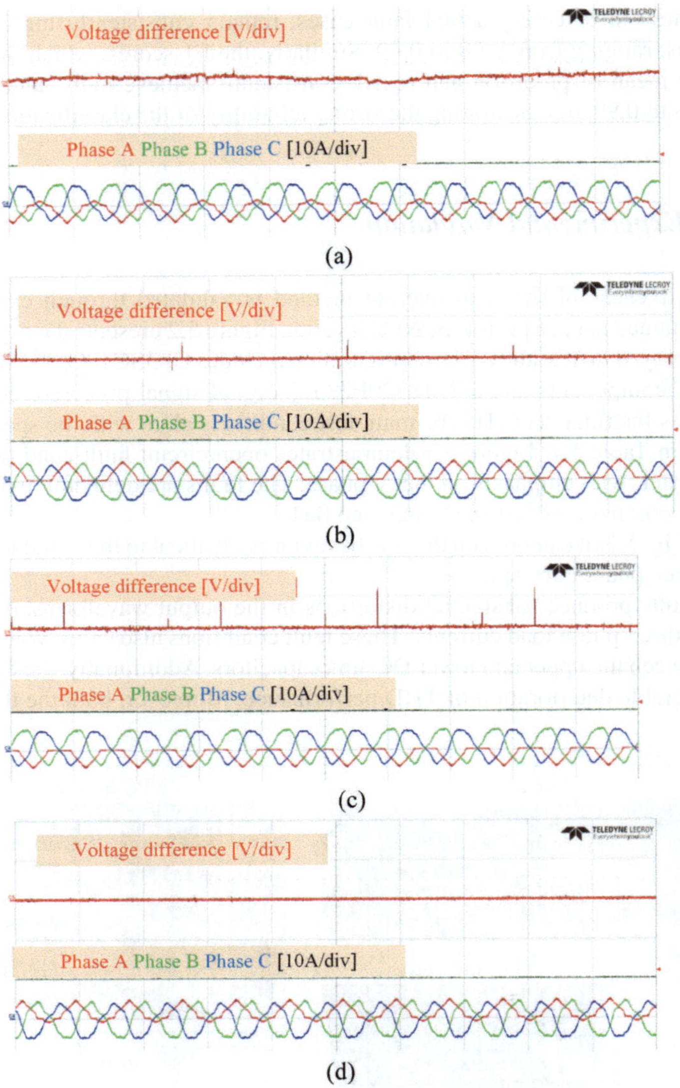

Fig. 5.3 Experimental three phase current waveforms across difference type of faults and normal case: **a** Sa1; **b** Sa2; **c** Sa3; **d** Sa4

current measurements and DC-link voltage references exhibit elevated noise levels compared to their simulated counterparts.

Table 5.5 summarizes the classification accuracy of various EML methods under different feature configurations for identifying open-circuit faults. Across all three scenarios, the RF and Soft (a voting ensemble of GaussianNB, RF, and AdaBoost) methods consistently achieve the highest accuracy of 0.97, demonstrating strong robustness and reliability in fault classification. When using only the three-phase currents as input features, the accuracy notably decreases. In this case, RF and the voting method reach the highest accuracy of 0.65, followed closely by Bagging at 0.64. These results highlight the importance of incorporating additional features to enhance detection performance. Specifically, when only the DC-link voltage difference and time are used, Bagging and AdaBoost yield the best results, with accuracies of 0.73 and 0.72, respectively. This suggests that these methods effectively utilize the temporal and voltage-based information for fault identification. In contrast, Gradient Tree Boosting and LightGBM consistently demonstrate lower accuracy across all scenarios, indicating limitations in extracting discriminative features for fault detection from the given inputs.

Table 5.6 presents the precision scores of various EML methods for detecting open-circuit faults in an NPC inverter. Among all fault types, RF, Bagging, AdaBoost, and the voting ensemble method consistently demonstrate strong performance, with precision scores ranging from 0.89 to 0.99. These results reflect a high level of accuracy in correctly identifying fault conditions. In contrast, Gradient Tree Boosting and LightGBM yield comparatively lower precision scores, suggesting limitations in accurately classifying certain fault types within the NPC inverter system. XGBoost achieves moderate precision across most fault types, indicating reasonable fault detection capability, though still trailing RF, Bagging, AdaBoost, and the ensemble method combining GaussianNB, RF, and AdaBoost. All methods generally perform well in recognizing normal operating conditions, achieving precision scores of 0.97 or higher, which demonstrates strong reliability in distinguishing non-fault scenarios.

Table 5.7 shows the RF algorithm results for identifying open-circuit faults in NPC inverters through precision, recall, and F1-score evaluation. Under normal

Table 5.5 Experimental results for the accuracy of ensemble machine learning methods

Methods	Accuracy-based three phase currents, voltage difference and time	Accuracy-based three phase currents	Accuracy-based voltage difference and time
RF	0.97	0.65	0.67
Bagging	0.96	0.64	0.73
AdaBoost	0.93	0.61	0.72
Gradient tree boosting	0.70	0.33	0.38
Lightgpm	0.85	0.46	0.42
Xgboost	0.90	0.50	0.45
Soft (GaussianNP + RF + Adaboost)	0.97	0.65	0.66

Table 5.6 Experimental results for the precision of ensemble machine learning methods for each type of open circuit fault

Fault type	Ensemble machine learning methods						
	RF	Bagging	AdaBoost	Gradient tree boosting	Lightgpm	Xgboost	GaussianNP + RF + Adaboost
Normal	0.99	0.99	0.99	0.97	0.98	0.99	0.99
A-phase S_1	0.96	0.96	0.93	0.59	0.84	0.88	0.96
A-phase S_2	0.96	0.96	0.92	0.65	0.80	0.86	0.96
A-phase S_3	0.98	0.97	0.96	0.77	0.91	0.96	0.98
A-phase S_4	0.98	0.97	0.95	0.74	0.88	0.93	0.98
B-phase S_1	0.95	0.93	0.89	0.55	0.79	0.82	0.97
B-phase S_2	0.96	0.95	0.90	0.61	0.77	0.82	0.90
B-phase S_3	0.97	0.98	0.96	0.78	0.92	0.95	0.97
B-phase S_4	0.97	0.97	0.95	0.75	0.87	0.92	0.98
C-phase S_1	0.96	0.95	0.91	0.58	0.77	0.83	0.96
C-phase S_2	0.97	0.96	0.91	0.65	0.77	0.84	0.97
C-phase S_3	0.98	0.97	0.95	0.80	0.91	0.94	0.97
C-phase S_4	0.98	0.97	0.95	0.77	0.88	0.93	0.97

operational states, RF achieves a precision value of 0.99. Across various fault categories, RF maintains robust precision levels, scoring between 0.95 and 0.99, demonstrating dependable diagnostic capability. The recall metrics, indicating the algorithm's capacity to detect actual fault instances, consistently record high values from 0.95 to 1.00. F1-scores, which balance precision and recall measurements, similarly show excellent performance within the same range. These results validate RF's comprehensive proficiency in identifying open-circuit malfunctions within NPC inverter architectures.

Table 5.7 Experimental results for the performance of the random forest method for each type of open circuit fault

Fault type	Precision	Recall	F1-score
Normal	0.99	1.00	1.00
A-phase S_1	0.96	0.97	0.96
A-phase S_2	0.96	0.96	0.96
A-phase S_3	0.98	0.98	0.98
A-phase S_4	0.98	0.98	0.98
B-phase S_1	0.95	0.95	0.95
B-phase S_2	0.96	0.95	0.95
B-phase S_3	0.97	0.98	0.97
B-phase S_4	0.97	0.98	0.98
C-phase S_1	0.96	0.96	0.96
C-phase S_2	0.97	0.96	0.96
C-phase S_3	0.98	0.98	0.98
C-phase S_4	0.98	0.97	0.98

5.3 Explainable Machine Learning

Local Interpretable Model-Agnostic Explanations (LIME) and Shapley Additive Explanations (SHAP) are advanced techniques that enhance the interpretability of machine learning models, enabling the identification of pivotal features for robust open fault detection, particularly in NPC inverter systems. These methodologies significantly contribute to a deeper understanding of fault detection mechanisms, thereby supporting the development of more reliable and efficient fault detection frameworks for power electronics applications.

The SHAP method, grounded in the principles of Shapley values from cooperative game theory [3, 4], systematically evaluates all possible feature combinations to quantify the contribution of each feature to the model's predictions. This is achieved by measuring the change in prediction outcomes when a feature is included or excluded across all feature subsets, subsequently averaging these contributions to derive importance values. These values provide a precise indication of each feature's impact on the final prediction.

In contrast, LIME is a model-agnostic interpretability technique that approximates complex, black-box machine learning models with simpler, interpretable local models, such as linear models or decision trees, to explain individual predictions [5]. By generating a local model fitted to data points proximate to the instance under consideration, LIME delivers observation-specific explanations. Its versatility allows it to be applied to any classifier, independent of the underlying algorithm, thereby offering flexible and interpretable insights into the decision-making process of complex models.

5.3.1 Simulation Validation

Figure 5.4a depicts a SHAP summary plot, which outlines the influence of individual features on the model's output for classifying open fault types in neutral-point-clamped (NPC) inverters. The plot indicates that features 2, 3, 4, and 6—corresponding to the three-phase currents and theta—exhibit relatively low importance in the classification of open fault types. Conversely, features 7, 1, 8, and 5, which represent the DC-link voltage difference, DC-link voltage (top and bottom), and time, respectively, are identified as highly influential features. The analysis highlights a key limitation: distinguishing between various open fault types using solely three-phase currents is challenging, as these rely on a single value per sampling time. To address this constraint, recent studies propose alternative methodologies, such as representing input signals as images or extended data sequences [3–5], to enhance fault detection accuracy and robustness. However, these approaches may not be applicable to real-time systems due to the large amount of input data required by machine learning methods. Figure 5.4b–f present SHAP summary plots for a specific open fault type in neutral-point-clamped (NPC) inverters, strengthening the critical roles of features 7, 1, 8, and 5—corresponding to the DC-link voltage difference, DC-link voltage (top and bottom), and time, respectively—in the classification of this fault type. These plots highlight the significant contributions of these features to accurate fault identification. Additionally, Fig. 5.5 comprises three graphs that provide essential insights for detecting open fault types using LIME. In Fig. 5.5a, the left graph displays a confidence interval, indicating a 100% accuracy in classifying the data as class 4, which corresponds to a fault in the A-phase S4 switch. The center graph illustrates the feature importance scores for this specific sample, with feature 7 (DC-link voltage difference) contributing a 41% importance score, followed by feature 8 (DC-link voltage, top and bottom) with a 24% score, and feature 5 (time) with a 21% score. The right graph highlights the top five features and their respective values, where features marked in purple support classification toward class 4, while those in other colors contribute to alternative classes. Consistent with these findings, Fig. 5.5b further confirms that the DC-link voltage difference, DC-link voltage (top and bottom), and time are pivotal in detecting open-circuit fault types, emphasizing their critical influence on fault classification accuracy.

5.4 Convolutional Neural Networks

5.4.1 Mathematical Model

CNNs are employed for feature extraction and fault classification to detect faults in various types of inverters [6, 7]. The mathematics of CNNs is explained in Chap. 2.

CNNs are used as classification algorithms to predict faults in NPC inverters. CNNs process normalized three-phase current data. The normalization step divides

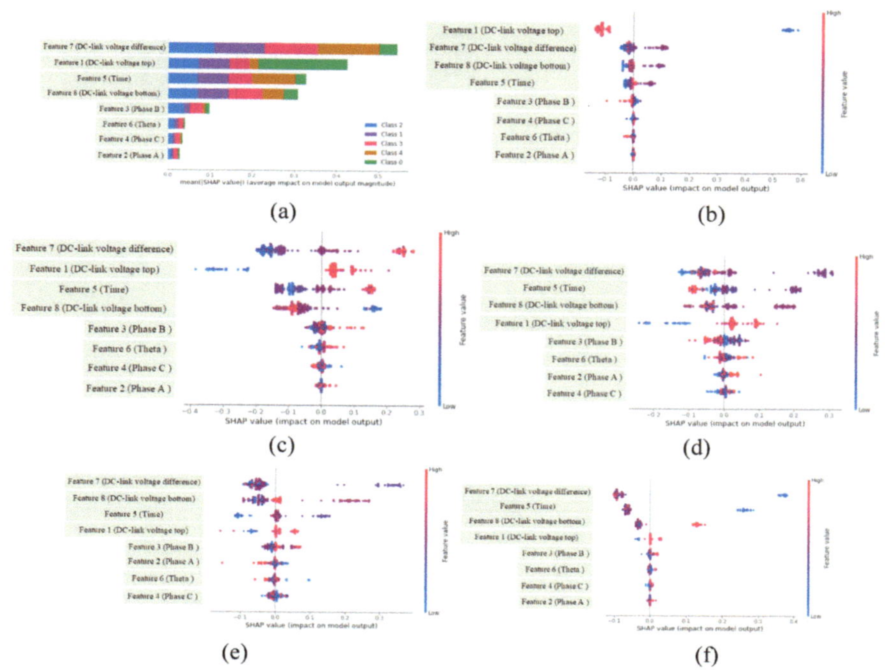

Fig. 5.4 SHAP summary plots: **a** all cases **b** case 0; **c** case 1; **d** case 2; **e** case 3; **f** case 4

current data by the respective current level for each simulation and experiment, resulting in a 3 × N matrix, where N represents the number of current data points in one cycle. The preprocessed data serves as input to the CNN, which is designed to detect 12 different open-switch fault cases in a three-phase circuit with four switches (S1 to S4), along with one label for normal operation, making a total of 13 labels. The CNN architecture consists of convolutional layers with 3 × 1 kernels, followed by max pooling layers that reduce dimensionality and emphasize significant features. These convolutional layers capture local patterns and extract complex features from the input data. Subsequently, a fully connected layer aggregates these features, and dropout is applied to prevent overfitting.

During training, back-propagation, with cross-entropy as the loss function, optimizes the network by adjusting weights to minimize the error between predicted and actual labels. An early stopping method monitors the validation loss after each epoch. It halts training if the loss fails to decrease for a specified number of epochs, thus retaining the weights from the best-performing epoch. This prevents overfitting and ensures that the model generalizes well to new data. During inference, the CNN processes normalized current data and outputs a probability distribution over 13 classes. The class with the highest probability indicates the predicted state of the NPC inverter, accurately identifying whether it is operating normally or if a specific switch is faulty. This robust classification mechanism ensures reliable fault detection

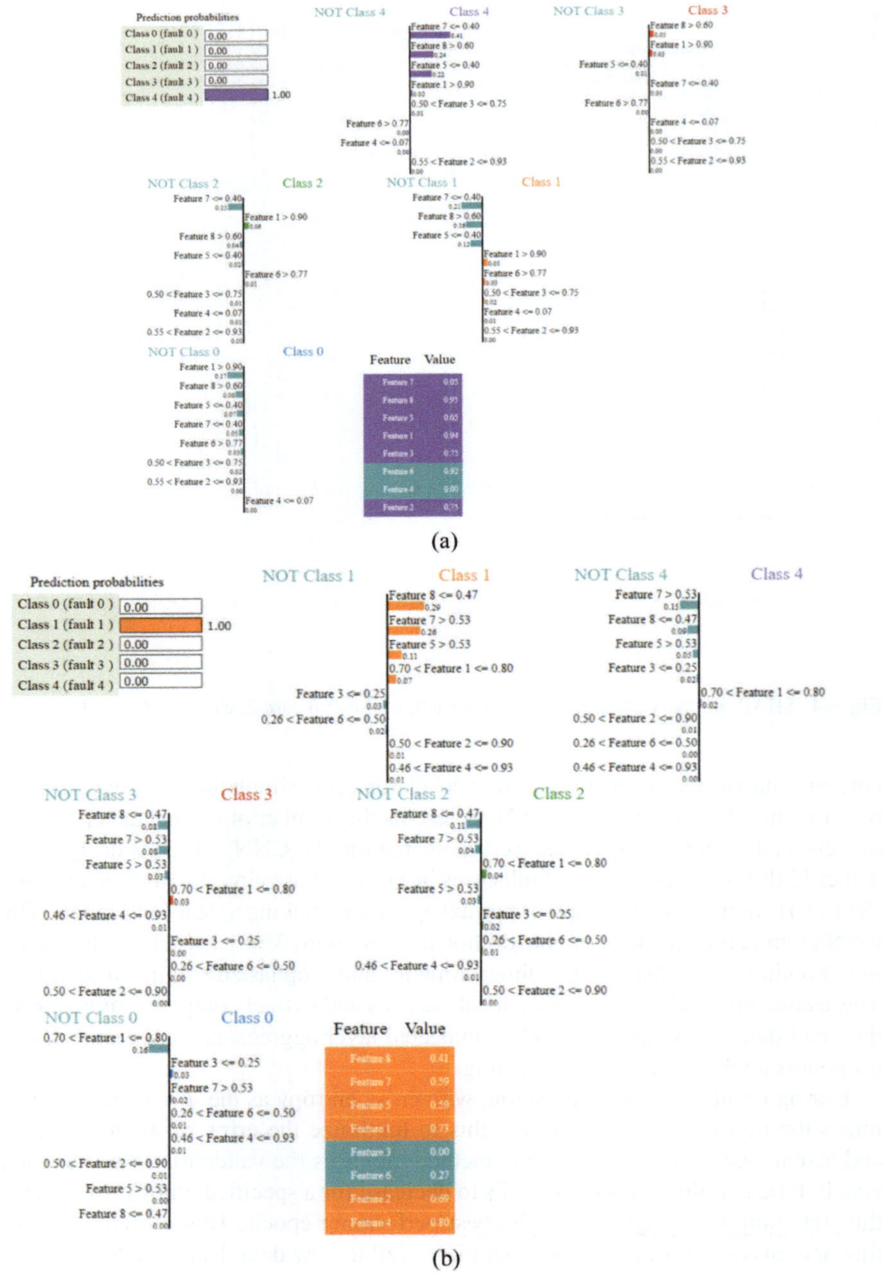

Fig. 5.5 LIME plots with different random samples: **a** case 4; **b** case 1

and efficient operation of the NPC inverter system. 1D CNNs have the following advantages: including require limited data compared to 2D CNNs, efficient training, cost-effective hardware implementation, and elimination of handcrafted features.

5.4.2 Simulation Validation

This study used the PSIM simulation tool to gather data for the CNN method. The simulation parameters are listed in Table 5.1. The proposed method was implemented in Python with machine learning libraries, such as Tensorflow, Keras, and PyTorch, and the parameters were tuned for optimal performance. The learning data were generated by conducting experiments at various current levels ranging from 5 to 15 A. This process was repeated for each current level to determine the number of cycles required for normal operation or a specific switch failure. The training-to-test data ratio was 8:2.

Table 5.8 lists the performance metrics (precision, recall, and F1-score) of the CNN classification model across various types of faults for the simulation dataset. The model demonstrates strong overall performance with high precision, recall, and F1-score values for most fault types. Precision indicates the accuracy of positive predictions, whereas recall measures the ability of the model to capture all positive instances. The F1-score represents a balance between precision and recall. The results demonstrated effective overall model performance. For normal or no faults cases, the model exhibited a high precision of 0.97, indicating a low rate of false positives. It also demonstrated an excellent recall of 0.99, by capturing the majority of instances correctly. An F1-score of 0.98 proposes a well-balanced performance. The model performed exceptionally well in identifying A-phase faults. A-phase S1 exhibited a good precision of 0.98 and a recall of 1.00, yielding an F1-score of 0.99. In addition, A-phase S2 exhibited performance similar to that of A-phase S1. A-Phase S3 and A-phase S4 demonstrated lower performances when compared to the performance of A-phase S1 and S2 with F1-scores of 0.95 and 0.94, respectively. Furthermore, the model exhibited excellent performance in identifying B-phase and C-phase faults. The majority of faults had an F1-score of 0.99. The minimum F1-score was 0.97 for S1 of B-phase and C-phase.

5.5 Machine Learning Methods for Fault Detection in DSPs

Fault detection in power electronics and motor drives is a complex and critical task. It is often complicated by the presence of various faults that can occur under diverse operating conditions. ML algorithms have been explored as potential solutions that offer advanced methods for various types of fault detection [7, 8]. However, many of these ML models, especially deep learning architectures such as CNNs, have a significant computational demand and involve numerous layers and operations that

Table 5.8 Simulation results for CNN-based methods

Fault type	Precision	Recall	F1-score
Normal	0.97	0.99	0.98
A-phase S_1	0.98	1.00	0.99
A-phase S_2	1.00	0.97	0.99
A-phase S_3	0.95	0.95	0.95
A-phase S_4	0.92	0.95	0.94
B-phase S_1	0.98	0.95	0.97
B-phase S_2	1.00	0.98	0.99
B-phase S_3	1.00	0.97	0.99
B-phase S_4	0.98	1.00	0.99
C-phase S_1	0.97	0.96	0.97
C-phase S_2	1.00	0.97	0.99
C-phase S_3	0.99	1.00	0.99
C-phase S_4	0.99	0.99	0.99

require extensive processing time. This high computational cost becomes particularly problematic in real-time applications, where fast response times are crucial to prevent system damage or failure. Additionally, the training and inference phases of such models can be resource-intensive, demanding considerable memory and computational power which may not be available in many of the embedded systems commonly used in power electronics and motor drive environments. Consequently, the integration of complex ML algorithms for fault detection remains a challenge, because the trade-off between accuracy and real-time performance has yet to be fully addressed. This section presents various strategies that successfully implement ML in DSPs for fault detection applications.

In [9], implementing CNNs on DSPs is challenging due to the high computational load required to process image-based input data for fault classification. To enable efficient ML integration on DSPs, reducing the input data is crucial. A recent study addressed this by proposing a low number of inputs data for fault detection in motor drive systems using ANN. Data acquisition occurs at specific locations within the system, with sampling performed six times during one electric cycle to capture detailed information including up to the second harmonic component of the electric angle. This stacked data, which includes voltage, current, speed, and torque, is used as input for the ML model. Faults, such as current-sensor and inverter switch-open faults, are identified by calculating errors between the reference values and the actual readings in the synchronous reference frame (SRF). Preprocessing steps, such as applying a low pass filter (LPF) with a cut-off frequency of 7500 Hz, are implemented to reduce noise and improve signal integrity. Data normalization is critical since voltage, current, speed, and torque variables are on different scales and must be adjusted to avoid having a negative impact on ANN performance.

In the deployment phase, the study focuses on optimizing the ANN for efficient execution on a DSP. The ANN is structured with two layers of 20 nodes. This is done

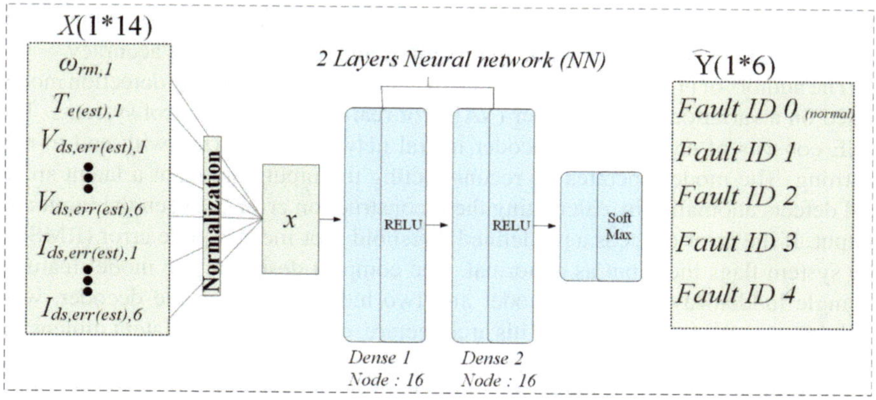

Fig. 5.6 ANN for fault detection

in accordance with Cybenko's theorem to ensure that it can approximate complex functions. The operating points of the motor are varied during data collection to capture both steady-state and dynamic conditions, while the rotor speed and torque are sampled every 0.1 s to capture dynamic behaviors. To minimize computational load, the ANN size is optimized without affecting performance, and the trained model parameters, such as weights and biases, are extracted for implementation on a TI28379S DSP. The ANN is written in C language and successfully deployed on the DSP, demonstrating that the system can effectively classify faults in real-time while maintaining computational efficiency. Figure 5.6 shows the proposed ANN method for fault detection.

The authors of [10] proposed a method for diagnosing current sensor faults in inverters using an ANN that addresses the challenges posed by out-of-distribution (OOD) data. OOD data refers to input data that differs significantly from the distribution of the training data used to develop a ML model. This discrepancy can result in suboptimal performance or unreliable predictions, because the model is unprepared to effectively handle these unfamiliar or unexpected inputs. The ANN is designed to classify current sensor faults based on an in-distribution dataset, which includes fault conditions with varied scale and offset errors. To make the ANN robust to OOD data, two key strategies are implemented: the introduction of a random noise dataset during training and the application of temperature scaling in the inference stage, which adjusts the prediction confidence without significantly increasing computational complexity. These techniques help improve the performance of the model when encountering unexpected data that could otherwise lead to misclassification. MATLAB/Simulink is used to simulate OOD data, including switch-open faults and random noise, to ensure the model can handle a variety of fault conditions. The ANN model, with 14 input nodes and two hidden layers of 16 nodes each, achieves a classification accuracy of 98.94% on the in-distribution dataset and significantly improves the area under the receiver operating characteristic (AUROC) score from

0.9170 to 0.9866 for OOD data. The results demonstrate the robustness of the model in handling unexpected data while maintaining high fault diagnosis accuracy.

The authors of [11] proposed a method that introduces an anomaly detection model based on a variational autoencoder (VAE) for real-time motor control systems. The VAE consists of encoder and decoder neural networks optimized with end-to-end learning. The model operates by reconstructing the input data from a latent space and detects anomalies by calculating the reconstruction error between the input and output. If the error exceeds a predefined threshold root-mean-square error (RMSE), the system flags the input as abnormal. The compact design of the model features a single hidden layer in the encoder and two hidden layers in the decoder, with each layer containing 12 nodes. This architecture, combined with a latent dimension size of two, allows for real-time processing on a TMS320F28379S DSP. For data acquisition, normal data is collected during both steady and dynamic states of motor operation. Meanwhile, abnormal data is gathered by inducing current-sensor and switch-open faults. The input data, consisting of voltage and current errors in the rotor SRF, is sampled six times per electric cycle at 60-degree intervals. The model is trained using only normal data, and its performance is evaluated based on the RMSE between normal and abnormal datasets. It can be concluded that the current errors along the d-axis and q-axis are the most effective input features, achieving an AUROC of 0.9873 in anomaly detection.

The authors of [12] introduced a novel multi-module neural network (MMNN) architecture for fault diagnosis in power systems. Unlike traditional approaches that rely on a single NN to diagnose faults across all six switches, the MMNN employs six specialized modules, with each module dedicated to diagnosing faults in a specific switch. By having separate modules for each switch, the system can independently diagnose individual switch faults and compound faults without increasing the complexity of the model or requiring additional features. Each module performs binary classification between normal operation and an open-fault condition, keeping the network architecture simple with just one hidden layer. The three inputs to each module represent the maximum current values obtained from the shunt resistor while injecting three space vector voltages. The output layer contains one neuron that indicates the fault probability for the respective switch. The MMNN design maintains computational simplicity by using the ReLU as the activation function for the hidden layer and the sigmoid function after the output layer to determine fault probability. Figure 5.7 illustrates the structure of the NN module responsible for diagnosing AT open faults. The modules are trained using backpropagation and binary cross entropy loss. The ability of each module to focus on a single switch fault enables precise and efficient fault diagnosis for both individual and compound faults.

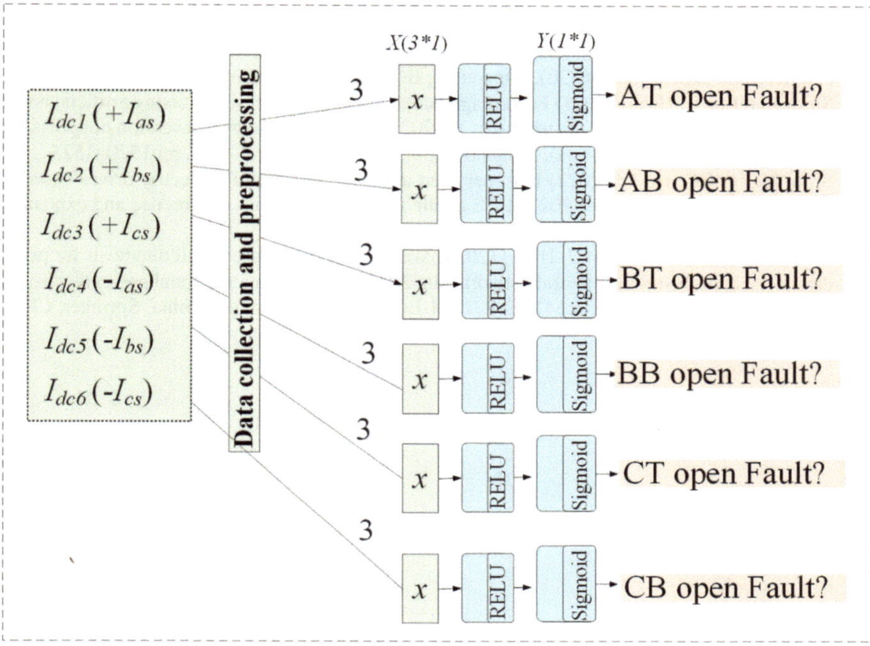

Fig. 5.7 Structure of the proposed multi-module neural network

References

1. Halabi LM, Alsofyani IM, Lee K-B (2021) Multi open/short circuit fault-tolerance using modified SVM technique for three-level HANPC converters. IEEE Trans Power Electron 36(12):13621–13633 (Springer, Cham)
2. Al-Kaf HAG, Lee JW, Lee KB (2024) Fault detection of NPC Inverter based on ensemble machine learning methods. J Electr Eng Technol 19:285–295 (Springer, Cham)
3. Lundberg SM, Lee SI (2017) A unified approach to interpreting model predictions. In: Proceedings of the 30th conference on neural information processing systems (NeurIPS), vol 30. Springer, Cham, pp 4765–4774
4. Lloyd S (1952) N-person games. In: Defense technical information center. Springer, Cham, pp 295–314
5. Al-Kaf HAG, Lee K-B (2023) Explainable machine learning method for open fault detection of NPC inverter using SHAP and LIME. In: 2023 IEEE conference on energy conversion (CENCON). Kuching, Malaysia, Springer, Cham, pp 14–19
6. Kiranyaz S, Gastli A, Ben-Brahim L, Al-Emadi N, Gabbouj M (2019) Real-time fault detection and identification for MMC using 1-D convolutional neural networks. IEEE Trans Ind Electron 66(11):8760–8771 (Springer, Cham)
7. Yuan W, Li Z, He Y, Cheng R, Lu L, Ruan Y (2022) Open-circuit fault diagnosis of npc inverter based on improved 1-D CNN network. IEEE Trans Instrum Meas 71:1–11 (Springer, Cham)
8. Kim S-H, Yoo D-Y, An S-W, Park Y-S, Lee J-W, Lee K-B (2020) Fault detection method using a convolution neural network for hybrid active neutral-point clamped inverters. IEEE Access **8**:140632–140642 (Springer, Cham)

9. Shim J-H, Lee J, Ha J-I (2021) Current-sensor and switch-open fault diagnosis based on discriminative machine learning model for PMSM driving system. In: 2021 IEEE energy conversion congress and exposition (ECCE). Vancouver, BC, Canada, Springer, Cham, pp 5098–5104
10. Shim J-H, Lee K, Ha J-I (2021) Fault diagnosis of inverter current sensor using artificial neural network considering out-of-distribution. In: 2021 IEEE 12th energy conversion congress and exposition—Asia (ECCE-Asia). Singapore, Singapore, Springer, Cham, pp 1520–1525
11. Shim J, Lim GC, Ha J-I (2022) Unsupervised anomaly detection for electric drives based on variational auto-encoder. In: 2022 IEEE applied power electronics conference and exposition (APEC). Houston, TX, USA, Springer, Cham, pp 1703–1708
12. Kim M, Kim J, Shin J, Shim J, Ha J-I (2024) Voltage injection-based fault diagnosis for power devices with multi-module neural networks. In: 2024 IEEE 10th international power electronics and motion control conference (IPEMC2024-ECCE Asia). Chengdu, China, Springer, Cham, pp 4660–4665

Chapter 6
Recommendations and Best Practices for ML in Real-Time Power Electronic Applications

Abstract This chapter builds on the previous discussion of implementing ML on DSPs for power electronics and motor drives, addressing key limitations such as computational constraints, data scarcity, and the complexity of deploying deep learning in real-time systems. To overcome these challenges, it explores emerging AI techniques like Physics-Informed Neural Networks (PINNs) and lightweight Convolutional Neural Networks (Light CNNs), which offer improved efficiency and feasibility for embedded applications. The chapter also addresses strategies to manage data limitations in ML-based power electronics, emphasizing the importance of robustness and uncertainty quantification for reliable real-time operation. It highlights methods for improving explainability in ML models to enhance trust and interpretability, crucial for industrial adoption. Experimental case studies are presented, showcasing real-time ML implementations across diverse embedded platforms, including FPGAs, DSPs, and hybrid systems. These implementations are critically analyzed to evaluate performance, feasibility, and practical deployment outcomes. Overall, the chapter identifies the most promising AI-driven approaches for enhancing real-time fault detection and control in power electronics and motor drive systems. It concludes by outlining future research directions focused on improving model efficiency, scalability, and trustworthiness, paving the way for next-generation intelligent control systems in industrial applications.

6.1 Future Research Directions in ML for Power Electronics

This section presents the future research directions in machine learning for power electronics, as summarized in Table 6.1.

H. A. G. Al-kaf and K.-B. Lee, *Machine Learning Integration in Power Electronics*,
https://doi.org/10.1007/978-981-95-3845-4_6

Table 6.1 Future research directions in machine learning for power electronics

Research direction	Description	Challenges addressed	Potential application
Surrogate modeling	Employ machine learning models (e.g., Artificial Neural Networks (ANN), Deep Belief Networks–Deep Neural Networks (DBN–DNN) to approximate complex and computationally expensive simulations such as Finite Element Method (FEM) models	Reduces computational burden while maintaining sufficient accuracy	Motor design optimization
Physics-Informed Machine Learning (PIML)	Incorporate prior physical knowledge and governing equations directly into the learning process of machine learning models	Addresses data scarcity and enhances physical realism absent in conventional ML models	Parameter estimation in power electronics converters
Uncertainty quantification in ML models	Utilize methods such as Bayesian Neural Networks (BNNs), Monte Carlo (MC) dropout, ensemble learning, and conformal prediction to quantify both aleatoric and epistemic uncertainty in model predictions	Enhances model reliability under noisy or unseen conditions	Fault diagnosis in electric vehicles (EVs) and aircraft systems
Explainable Artificial Intelligence (XAI)	Apply interpretability techniques such as SHAP, LIME, Individual Conditional Expectation (ICE), Grad-CAM, and DeepLIFT, along with theoretical metrics (e.g., Lipschitz continuity), to explain model decision-making	Mitigates the black-box nature of deep learning models, promoting stability, robustness, and trust in power electronic systems	Fault diagnosis
Data imbalance handling techniques	Implement techniques such as Safe-SMOTE to mitigate class imbalance in training data	Addresses biases, overfitting, misclassification, low recall, and unreliable model performance	Fault diagnosis
Handling data scarcity	Apply transfer learning to leverage knowledge from related domains, reducing the dependency on large labeled datasets	Overcomes the challenge of insufficient training data	Fault diagnosis
Data generation	Use generative models such as Generative Adversarial Networks (GANs) to synthesize realistic data samples	Alleviates the issue of data insufficiency in critical applications	Battery health estimation (State of Charge (SOC), State of Health (SOH)

(continued)

Table 6.1 (continued)

Research direction	Description	Challenges addressed	Potential application
Low-complexity models	Design optimized CNN architectures and deploy them using frameworks such as ONNX and TensorRT for real-time inference on edge devices (e.g., Jetson Xavier NX)	Minimizes model size and computational load without affecting performance	Real-time fault diagnosis
Real-time implementation of reinforcement learning	Modify the reinforcement learning framework by incorporating delayed observations and previously taken actions into the state representation. Additionally, integrate safety constraints directly into the reward function	Enables a robust and safe learning framework tailored for the dynamic requirements of electric drive systems	Electric drive control and optimization

6.1.1 Surrogate Model-Based Machine Learning

A surrogate model mimics system dynamics that are difficult to formulate or require intensive computation. In the iterative design process, ML-based surrogate models reduce computational effort by simplifying the relationship between design variables. They are used as substitutes for more computationally expensive methods, such as FEM to predict machine performance, including torque, torque ripple, and loss in the motor. The surrogate model-based ML approach has been implemented successfully in a DSP to find the optimal droop gain for DC microgrids onboard more electric aircraft [1]. The process involves generating data through simulations for various combinations of droop gains (k_{d1}, k_{d2}, and k_{d3}) and measuring the corresponding performance metrics, such as current sharing ratios ($n1$, $n2$) and normalized bus voltage (V_{bn}). These data are used to train a surrogate model (ANN), which models the relationship between droop gains and system performance. Once the surrogate model is trained, the design space is sampled using small steps, creating numerous design points. These points are then evaluated using the surrogate model, and a fitness function (f_{ann}) is applied to find the optimal droop gain settings. While the surrogate model is employed in this method, it still requires significant computational effort due to the large number of design points and fitness function evaluations. An additional improvement over previous methods is that the proposed approach utilizes reverse data training, allowing the ANN to directly predict the optimal droop gains based on input performance metrics, such as power sharing ratios and bus voltage [2]. Once trained, the surrogate model can instantly compute the optimal droop gains for any specified performance objective, eliminating the need for extensive design space sampling and fitness evaluations. This advancement further reduces computational complexity and enables real-time tuning of droop gain parameters, enhancing the overall efficiency and responsiveness of the control system. In [3], a surrogate model based on the deep belief network-deep neural network (DBN-DNN) algorithm replaces the motor control strategy and harmonic injection module, optimizing the stator voltage in the dq-axis and reducing the output torque ripple of a PMSM. The DBN consists of multiple layers of restricted Boltzmann machines (RBMs), with four layers identified as optimal to achieve the best results. The particle swarm optimization (PSO) algorithm determines the number of neurons in the hidden layers, set to (20, 20, 20, 20), with RMSE as the fitness function to guide the training process. This surrogate model based on the DBN-DNN predictive model is embedded within FOC, which simplifies the traditional double closed-loop structure of PI control.

6.1.2 Physics-Informed Machine Learning

ML in power electronics faces significant challenges due to the scarcity of labeled data. This data limitation often restricts the practical applicability and reliability of ML models in real-time applications. Furthermore, traditional ML methods are often unable to derive interpretable insights from vast amounts of data and typically lack the ability to incorporate physical principles, making their predictions less accurate and less reliable in domains where such constraints are essential. Physics-informed machine learning (PIML) has emerged as a promising approach to address these challenges by integrating prior physical knowledge directly into the learning process [4]. Although a relatively new paradigm, PIML has gained traction across various fields, including fluid mechanics, chemistry, geology, and power systems. It enhances model accuracy and interpretability by embedding physical laws as constraints during the training process. Several papers have documented the diverse applications of PIML across these domains, showcasing its versatility and growing importance. One of the core approaches within PIML is the physics-informed neural network (PINN). PINNs embed physical laws, often represented as partial differential equations (PDEs), into the neural network architecture. This is typically done by incorporating the physical equations into the loss function of the network, allowing the model to learn from data and satisfy physical constraints. For example, a PINN trained to model fluid dynamics minimizes prediction errors while ensuring its predictions adhere to the governing equations, like the Navier–Stokes equations. This approach enables PINNs to leverage unlabeled data more effectively, because the embedded physics provide guidance on the expected behavior, making PINNs particularly valuable when labeled data is limited.

A simple logistic equation is presented for modeling population growth, which describes how a population changes over time.

$$\begin{aligned} \frac{df(t)}{dt} &= Rt(1-t) \\ f(t=0) &= 1 \end{aligned} \tag{6.1}$$

To find a specific solution for this equation, an initial condition (sometimes called a boundary condition) is needed. A neural network is used to approximate *f(t)*, letting it learn how the population changes over time. This is advantages because it can model complex behaviors in growth patterns by training the network to follow the differential equation. To compute how the population f changes over time (i.e., the derivative), automatic differentiation is used. This feature is provided by deep learning libraries like PyTorch. Automatic differentiation makes it possible to easily compute the derivatives of functions defined by the neural network.

```
def df(f: simple_NN, x: torch.Tensor = None, order: int = 1) -> torch.Tensor:
    """Compute neural network derivative with respect to input features using PyTorch
autograd engine"""
    df_value = f(x)
    for _ in range(order):
        df_value = torch.autograd.grad(
            df_value,
            x,
            grad_outputs=torch.ones_like(x),
            create_graph=True,
            retain_graph=True,
        )[0]

    return df_value
```

The NN should satisfy the following equation by adding the following physics informed regularization term to the loss.

$$
\begin{aligned}
&\frac{dNN(t)}{dt} - Rt(1-t) = 0 \\
&Loss_{PDE} = \frac{1}{N}\sum_{i=1}^{N}\left(\frac{dN}{dt}\Big|_{t_i} - Rt_i(1-t_i)\right)^2 \\
&Loss_{BC} = (NN(t_0) - 1)^2 \\
&Loss_{Final} = Loss_{PDE} + Loss_{BC} + Loss_{data}
\end{aligned}
\tag{6.2}
$$

where t_i are collocation points, i.e., a set of points from the domain where the differential equation is evaluated.

Then a loss function is computed that defines a custom final loss for training a neural network to approximate population growth according to specific guidelines. It combines three main components: PDE loss, boundary condition loss (BC), and MSE loss ($Loss_{data}$). The PDE loss ensures the predictions of the model align with a known growth rule, penalizing any deviation from this expected behavior. The boundary condition loss enforces an initial population value, guiding the model to start at a specific known value, while the MSE loss minimizes differences between the predictions of the model and actual data points. By combining these losses, the network learns to generate predictions that follow the expected growth pattern, begin at the correct initial value, and fit any available training data.

Then the neural network is trained using Adam optimizer based on the code below.

```
# Wrap everything into a function
def compute_loss(nn: simple_NN,
                 t: torch.Tensor = None,
                 x: torch.Tensor = None,
                 y: torch.Tensor = None,
                 ) -> torch.float:
    """Compute the full loss function as pde loss + boundary loss
    This custom loss function is fully defined with differentiable tensors therefore
    the .backward() method can be applied to it
    """
    pde_loss = df(nn, t) - R * t * (1 - t)
    pde_loss = pde_loss.pow(2).mean()
    boundary = torch.Tensor([0.0])
    boundary.requires_grad = True
    bc_loss = nn(boundary) - ft0
    bc_loss = bc_loss.pow(2)
    mse_loss = torch.nn.MSELoss()(nn(x), y)
    tot_loss = pde_loss + bc_loss + mse_loss

    return tot_loss


model = simple_NN()
optimizer = torch.optim.Adam(model.parameters(), lr=1e-2)

# Train
for ep in range(2000):

    loss = compute_loss(model, t, x_train, y_train)

    # Backpropagation
    optimizer.zero_grad()
    loss.backward()
    optimizer.step()

    if ep % 200 == 0:
        print(f"epoch: {ep}, loss: {loss.item():>7f}")
```

PIML can be classified into three types, as presented by Xinze Li. These methods including physics-in-loss, physics-in-architecture, and physics-in-initialization. Each of them is designed to integrate physical principles, such as PDEs, into ML models to ensure solutions that align with the underlying physics. These approaches are particularly valuable in fields such as power electronics, where physical consistency is critical.

Physics-in-loss incorporates physical laws directly into the loss function of the model [5]. In this method, the neural network is trained to fit available data and to satisfy the governing PDEs by including penalty terms in the loss function that measure deviations from these equations. For power electronics, this could mean enforcing electrical principles or energy conservation, ensuring that predictions remain physically realistic even with limited or noisy data. This approach is versatile, adaptable to various network architectures, and particularly useful for solving PDEs and estimating system parameters. However, it requires careful balancing of data-driven and physics-based terms in the loss function, which can complicate implementation.

Physics-in-architecture neural network (PANN) embeds physical knowledge into the structure of the neural network itself [6]. PANN, a recurrent network, is designed to discretize PDEs using numerical methods, modeling the time-domain behavior of power electronics by predicting state variables based on prior states. This makes it inherently explainable, because its neural parameters correspond to physical converter parameters, and it captures phenomena like switching behaviors and operating modes. In power electronics, PANN stands out due to its ability to operate effectively with minimal data requirements, which requires far less training data, and its lightweight design. While highly effective and interpretable, crafting such an architecture demands significant domain expertise, and its specificity may limit its adaptability to entirely different systems.

Physics-in-initialization leverages physical principles to set the starting point of the model parameters or states before training. Rather than beginning with random weights, the network is initialized with values derived from approximate physical solutions or known system behaviors, guiding the learning process toward plausible outcomes. In power electronics, this might involve initializing a model with parameters from simplified circuit models or typical operating conditions, speeding up convergence and reducing the likelihood of unphysical results. This method is especially helpful for ill-posed problems where data alone is insufficient. However, its success depends heavily on the accuracy of the initial guess, and it may struggle if the starting conditions do a poor job of representing the true system.

6.1.3 Addressing Uncertainty in Machine Learning

Various ML methodologies have been developed to address the uncertainty of ML methods, including Bayesian inference, ensemble methods, Monte Carlo dropout, and conformal prediction. These approaches offer distinct strategies for managing uncertainty by accommodating both the inherent variability in data and limitations in model specification.

ML models, particularly deep learning-based approaches, encounter two primary sources of uncertainty. Aleatoric uncertainty, also known as data uncertainty, arises from inherent noise or variability in data and is irreducible. In power electronics, factors such as sensor noise or measurement inconsistencies contribute to aleatoric uncertainty, reducing prediction reliability. Meanwhile, epistemic uncertainty, or knowledge uncertainty, results from limited training data or incomplete knowledge of the problem domain. Epistemic uncertainty can be mitigated by gathering more data or enhancing the generalization capabilities of the model. In fault diagnosis, unseen fault conditions introduce high epistemic uncertainty, making it critical to employ uncertainty-aware models such as Bayesian neural networks (BNNs) [7]. BNNs provide a powerful approach to fault diagnosis in motor drives by modeling uncertainty in predictions. Unlike traditional neural networks that use fixed weights, BNNs treat network parameters as probability distributions, allowing for improved robustness and interpretability. Conventional fault diagnosis methods often fail to account for noise and unseen fault conditions, leading to unreliable predictions. BNNs address this issue by providing confidence estimates for each diagnosis, making decision-making more reliable.

Ensemble methods enhance prediction reliability by aggregating outputs from multiple models rather than relying on a single predictor [8]. This approach involves training diverse models with different initial conditions, architectures, or data subsets, and combining their outputs through techniques like averaging or majority voting. The variability among these models serves as an indicator of uncertainty: high agreement suggests confidence, while discrepancies signal greater uncertainty. Popular implementations, such as Random Forests and Deep Ensembles, are well-suited for power flow analysis in power electronics. However, the requirement to train and execute multiple models increases the computational costs, which can be a limitation in real-time scenarios.

Monte Carlo (MC) dropout is an uncertainty estimation technique that extends dropout regularization, originally designed to mitigate overfitting in neural networks, into the inference phase [9]. By randomly deactivating network neurons during multiple prediction iterations, MC dropout generates a range of outputs for the same input. The spread of these predictions reflects uncertainty. Narrow distributions indicate confidence, while wider distributions suggest ambiguity. MC dropout integrates seamlessly into existing neural architectures, making it a practical choice for real-time monitoring of inverter performance. However, its effectiveness depends on the assumption that dropout-induced variability approximates a true probabilistic model.

Conformal prediction is a powerful, distribution-agnostic framework designed to produce prediction sets with reliable uncertainty bounds, without relying on specific assumptions about the underlying data distribution. This approach works by assessing how well new data instance aligns with patterns observed in historical data, leveraging a concept called "conformity." Based on this evaluation, it constructs prediction intervals or sets that are guaranteed to contain the true outcome with a user-defined confidence level, such as 95%. Unlike traditional statistical methods that might assume normality or other distributional properties, the flexibility of this method makes it applicable across diverse datasets and domains. Conformal prediction is highly relevant for power electronics applications, such as aiding PV power suppliers in decision-making for the day-ahead electricity market [10].

The impact of noise on uncertainty estimation is a critical consideration. When the signal-to-noise ratio (SNR) decreases, aleatoric uncertainty becomes more pronounced, affecting model reliability. BNNs and other probabilistic methods effectively account for this noise, ensuring robust performance even in challenging environments. By integrating probabilistic assessments into AI-driven decision-making, these models enhance trustworthiness, providing interpretable and actionable insights for engineers working with AI-based fault diagnosis in power electronics.

6.1.4 Explainable Machine Learning in Power Electronics

ML has demonstrated significant potential in power electronics, offering advanced control strategies, fault diagnosis, and system optimization. However, a major challenge in applying ML to critical power electronics systems is its "black-box" nature. The complexity of deep learning models, often comprising numerous layers and thousands of parameters, makes their decision-making process difficult to interpret. This lack of transparency raises concerns about trust, reliability, and practical adoption in mission-critical applications such as electric aircraft, power grids, and renewable energy systems. To address this issue, explainable artificial intelligence (XAI) methods have emerged, aiming to improve interpretability while maintaining model performance.

XAI techniques are broadly categorized into local and global explanations. Local explanations focus on interpreting a single prediction by determining which input features contributed most to the output. In contrast, global explanations provide insights into how an entire model makes decisions across different inputs. Several XAI techniques have been proposed, including SHAP, Local LIME, and individual conditional expectation (ICE), which are widely used in power systems. SHAP assigns a contribution score to each input feature based on cooperative game theory, quantifying their influence on the output of the model. This is particularly useful for optimizing control strategies in power converters by identifying the most critical system parameters. LIME, on the other hand, perturbs the input data and approximates the decision-making of the ML model with a simpler, interpretable surrogate model. ICE is a technique used in the field of ML to visualize and interpret the conditional effects of a specific variable on the outcome of interest. It provides a way to understand how the relationship between a particular feature and the target variable changes across different values of other variables. ICE is particularly useful in cases where there is an interest in understanding the impact of a single variable in a complex model, especially in the context of non-linear relationships. ICE provides a more complete picture of the relationship than traditional global measures such as the average effect, because it determines the conditional expectation for each observation in the dataset if the variable of interest has different values. This helps engineers understand specific predictions, such as why a neural network-based inverter control system responds to voltage fluctuations in a certain way. In addition, the SHAP and individual conditional expectation (ICE) analyses provided a comprehensive understanding of model behavior, highlighting the critical influence of parameters such as reaction time, nominal power, and price elasticity on the stability of the decentralized smart grid [11].

For deep learning models in power electronics applications, techniques like Gradient-weighted Class Activation Mapping (Grad-CAM) and Deep Learning Important Features (DeepLIFT) provide visualization-based explanations. Grad-CAM highlights critical input regions that influence model decisions, which is beneficial for diagnosing faults in power converters using image-based thermal analysis. DeepLIFT assigns importance scores to neurons, offering insights into how specific layers contribute to decision-making, which is valuable in tuning deep reinforcement learning controllers for grid-connected inverters [12].

Beyond interpretability, mathematical explainability is crucial for ensuring stability and robustness in power electronics ML models. A key mathematical property used for this purpose is Lipschitz continuity, which ensures that small input changes do not lead to large, unpredictable variations in model output. This property is essential in real-time control applications, where instability can cause oscillations, overshoots, or even system failures. Additionally, Lipschitz-aware learning rate selection can enhance training convergence, which prevents divergence issues in AI-driven power management systems [13].

XAI methods can be integrated into ML models through two approaches: intrinsic (model-specific) and post-hoc (model-agnostic). Intrinsic XAI incorporates explainability directly into the architecture of the model, making it inherently interpretable. This approach is useful for designing new AI-driven controllers in power converters. In contrast, post-hoc XAI is applied after training and is model-independent, allowing for the analysis of already-deployed ML systems. This is particularly advantageous in power electronics, where existing AI-based models can be evaluated and refined without retraining them from scratch.

6.1.5 Strategies for Handling Data Challenges in ML-Based Power Electronics Applications

In power electronics applications, data issues pose significant challenges for machine learning models, including fault diagnosis. Fault occurrences are rare compared to normal operating conditions. This imbalance skews classification models, leading to high accuracy for normal states but poor fault detection performance. To mitigate this issue, an oversampling technique using the safe-level synthetic minority oversampling technique (Safe-SMOTE) is employed. Safe-SMOTE starts by identifying a minority instance and its k-nearest minority neighbors. Then it calculates a safe level for each of them based on how many minority neighbors surround it, where higher counts mean safer, denser areas. New samples are created by interpolating between the instance and a chosen neighbor, with the position adjusted by the safe level ratio. If the instance is in a safer spot, the synthetic point shifts closer to it, reducing the risk of generating noisy or misclassified samples near majority class boundaries. By generating synthetic fault data in reliable regions of the dataset, Safe-SMOTE ensures a balanced dataset without distorting essential system characteristics [14].

In power electronics, transferring diagnostic models between different systems presents another challenge. Differences in operating conditions and hardware variations often render pre-trained models ineffective in new environments. A transferable, data-driven fault diagnosis method addresses this issue by leveraging manifold feature learning and model adaptation techniques. The methodology involves training an initial fault classifier on source system data and adapting it to target system conditions using a maximum mean discrepancy (MMD) criterion. By minimizing the distribution divergence between datasets, the adapted model generalizes effectively to new systems without requiring extensive retraining. An extreme learning machine (ELM) classifier is employed due to its fast learning speed and computational efficiency, making it suitable for real-time fault diagnosis [15].

Energy storage systems, particularly lithium-ion batteries, face challenges in estimating state-of-charge (SOC) and state-of-health (SOH) due to limited and costly data availability. Experimental trials for collecting comprehensive battery datasets are time-consuming and expensive. Generative adversarial networks (GANs) offer a promising solution for synthetic time-series data generation, augmenting existing datasets and improving machine learning-based estimation models. A time-series GAN captures voltage, current, temperature, and SOC trends, ensuring the generation of high-fidelity synthetic data. The generator-discriminator framework, which is enhanced by a supervised loss component, refines data generation by preserving both feature distributions and temporal dependencies. This approach facilitates robust battery modeling and enhances anomaly detection capabilities in power electronic systems [16].

6.2 Experimental Implementation of Machine Learning Algorithms

6.2.1 Implementation of Lightweight CNNs on Jetson Xavier NX

CNNs are widely employed for fault detection in complex electrical systems, including multilevel inverters. Traditionally, CNN-based fault diagnostics are performed offline, where data is collected and analyzed post-event. However, this

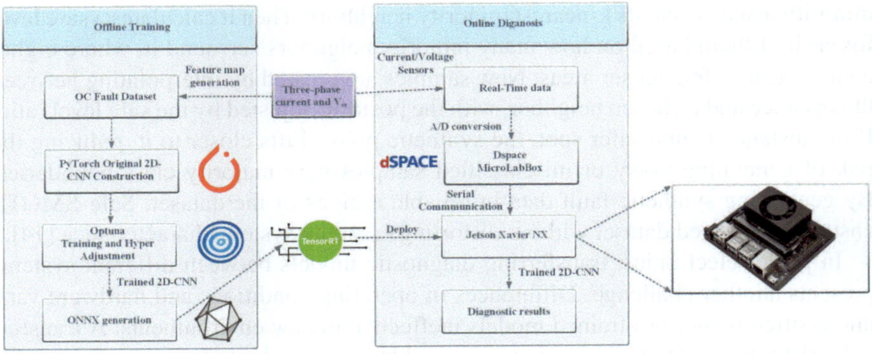

Fig. 6.1 Flowchart of the edge-based fault diagnosis method

approach imposes significant limitations on real-time monitoring, where immediate fault detection is essential to ensure system reliability and operational efficiency.

The authors of [17] focused on achieving real-time fault detection, specifically targeting open-circuit (OC) faults in a three-level active neutral-point-clamped (3L-ANPC) inverters, as illustrated in Fig. 6.1. To enable real-time diagnostics, a trained CNN is deployed on an edge computing device and optimized using TensorRT. This optimization process enhances computational efficiency by reducing model complexity and inference time, making it suitable for rapid fault detection. The implementation process begins with the training and optimization of the CNN. Initially, the model undergoes offline training to learn characteristic fault patterns. During this phase, an automated hyperparameter tuning framework (Optuna) is employed to optimize the model parameters, thereby improving classification accuracy and generalization to previously unseen data. This ensures that the CNN remains robust under varying fault conditions.

Following the training phase, the model is prepared for deployment. To facilitate compatibility across different hardware platforms, the trained CNN is first converted into the open neural network exchange (ONNX) format. Subsequently, TensorRT is utilized to further optimize the model by reducing its size and computational complexity. The optimized CNN is then deployed on Jetson Xavier NX, a high-performance edge computing platform capable of supporting real-time diagnostics.

To improve fault detection accuracy, a threshold-based decision mechanism is integrated into the system. This mechanism serves as a safeguard against false positives and false negatives, enhancing the reliability of CNN predictions. The optimal threshold values are determined through a confusion matrix analysis, ensuring a balance between high detection accuracy and rapid inference speed.

A comparative analysis is conducted to assess the efficiency of the optimized CNN against its original version. The results indicate that the lightweight CNN significantly reduces computational resource demands while preserving high classification accuracy. When executed on a GPU-based edge device such as the Jetson

Xavier NX, it achieves superior processing speeds and cost-effectiveness compared to alternative hardware platforms, including FPGA-based implementations.

The findings demonstrate that CNN can accurately and efficiently diagnose faults in real time, demonstrating its suitability for the real-time health monitoring of inverter systems.

6.2.2 Implementation of Reinforcement Learning for Motor Drives

To apply reinforcement learning (RL) in real-time motor control experiments, several modifications were made to the RL framework to address the unique demands of electric drive systems [18]. Implementing RL faces challenges such as real-time processing requirements, safety constraints, system delays, and the limitations of embedded hardware.

One major adaptation involved redefining the RL experience tuples to account for system delays. Traditionally, an experience in RL consists of the state, action, reward, and next state. However, in motor control, there is a delay before an action takes effect on the motor. To handle this, the experience tuple was expanded to include not just the immediate next observation but also the observation after the effect of the action becomes visible. By including both the reward and the delayed observation, the modified experience tuple provides the RL agent with feedback that reflects the actual system response over time. This adaptation helps the agent learn more accurate cause-and-effect relationships in the behavior of the motor.

Another critical modification was the inclusion of an action feedback mechanism. Due to digital control delays, the effects of an action are not immediately observable in the current timestep. To compensate for this, the previously taken action is appended to the current observation and fed into the RL model. This approach allows the agent to take its last applied action into account when estimating the next system state, which helps it learn to anticipate and mitigate overshoot or other dynamic effects. Essentially, this setup enables the RL model to better manage system inertia and delay, enhancing stability and precision in real-world control.

Ensuring motor safety was a top priority, especially to prevent overcurrent events that could damage both the motor and associated power electronics. To address this, safety constraints were incorporated directly into the reward function. The reward function penalizes actions that exceed a predefined current limit, thus encouraging the agent to respect these constraints while pursuing its control objectives. The reward function is specifically shaped to prioritize control accuracy, which is represented by the tracking error, while also incorporating a penalty term for constraint violations. This design helps the agent learn safe and effective control practices.

To bridge the gap between simulation and real-world testing, a rapid control prototyping toolchain was developed. This setup includes software-in-the-loop (SIL) simulations, allowing algorithms to be tested in a controlled environment before being transferred to actual hardware. The RL models were pre-trained in simulation, using a custom setup with dSPACE hardware, which enabled a smoother transition to the physical motor setup. On the hardware side, commercial off-the-shelf embedded systems such as dSPACE MicroLabBox and DS1006MC were used to run the control algorithms in real time. These systems allowed the RL agent to function within the strict timing requirements of motor control applications while facilitating easy modifications to RL algorithm parameters as needed.

To improve the training stability and effectiveness of the RL controller, exploration noise (used to encourage the agent to try different actions) was applied and carefully tuned. Various configurations of Ornstein–Uhlenbeck (OU) noise were tested to find the most effective balance between exploration and control precision. Additionally, the actor and critic neural networks were simplified to fit within the resource constraints of the embedded hardware. Initial tests showed that smaller network structures were more stable in real-world setups, because they can meet the timing requirements without compromising learning capacity.

6.2.3 Integrated DSP and FPGA for Deep Learning

A hybrid integration of DSP and FPGA is implemented for the deep-learning-aided closed-loop synchronous rectification (SR) strategy to address challenges in capacitor–inductor–inductor–capacitor (CLLC) converters, as illustrated in Fig. 6.2 [19]. Traditional SR strategies for such converters often use open-loop methods that rely on either diode on-state detection or model-based calculations. The former requires high-speed, complex sensors, while the latter only works under specific operating conditions. This deep-learning model provides initial SR signals, while temperature-based closed-loop optimization using gradient descent further fine-tunes the SR phase shifts to minimize the conduction loss and heat-sink temperature. Optimization is performed on the heat-sink temperature which is measured by thermocouples. This helps counter discrepancies between ideal and actual circuit parameters, ensuring ideal SR conditions. The neural network is a fully connected FNN with four layers: an input layer, two hidden layers (each containing 32 neurons), and an output layer. The FNN uses the Swish activation function for non-linear transformations, enabling it to accurately model complex relationships between SR phase shifts and the operating conditions of the CLLC converter. To generate a reliable and generalized model, the network is trained with an extensive dataset representing different operating conditions.

Training is performed using TensorFlow and Keras frameworks, with parameters initialized using zero and uniform random values. Data is normalized to a [0, 1] range for optimal training performance, and the Adam optimizer is used to minimize MSE between the model predictions and actual data, achieving a final MSE of 1.04×10^{-6} after 3775 epochs. Real-time deployment of the model is facilitated by its compact architecture, allowing it to process data rapidly and accurately within the control loop of the CLLC converter. The experimental setup for testing the deep-learning-aided

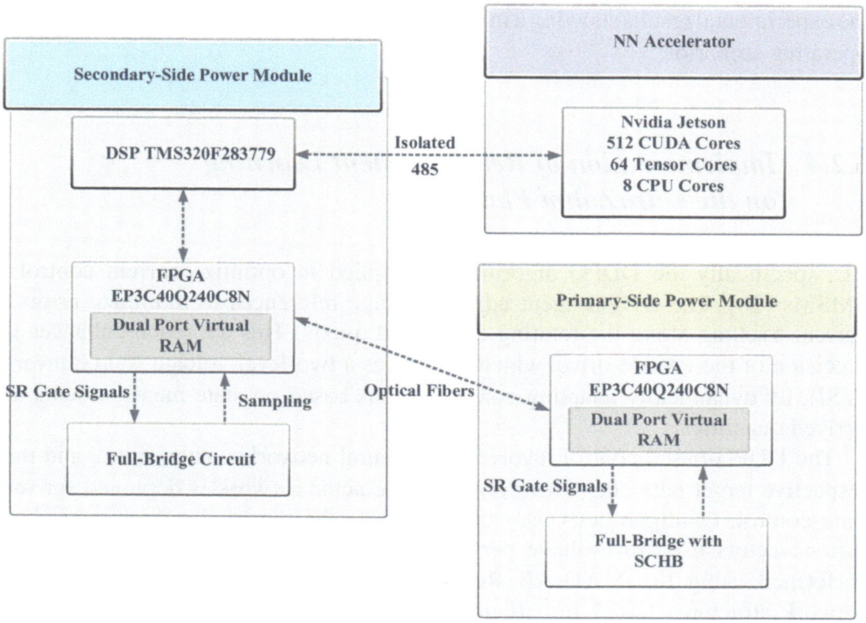

Fig. 6.2 Interactive block diagram of the control hardware architecture integrating DSP and FPGA for deep-learning execution

SR strategy uses a 400 V/300 V, 2.4 kW CLLC converter prototype with silicon carbide (SiC) components to verify the efficiency of the proposed method across various conditions.

In this setup, two field-programmable gate arrays (FPGAs) are employed to manage the PWM outputs and data sampling on both the high-voltage and low-voltage sides of the converter. The FPGAs work in tandem with a digital signal processor (DSP), which handles critical functions like timing control, proportional-integral (PI) control, PWM modulation, and delivering sampled load conditions to the Nvidia Jetson neural network accelerator. The DSP and FPGAs communicate via an isolated optical fiber link to maintain galvanic isolation, ensuring safety and fail-safe operation.

The Nvidia Jetson AGX Xavier, which serves as a neural network accelerator, processes the deep-learning model that predicts SR signals, which are further optimized in real-time based on temperature feedback. The high-speed computation capabilities of the Jetson (processing each SR model in just 2 μs with the BLAS library) make it possible to meet real-time demands while retaining flexibility for development and troubleshooting.

Finally, the data exchange between the DSP and FPGAs is enabled by the virtual dual-port RAM in the FPGAs, allowing smooth communication. This combined control and optimization setup ensures the SR strategy can operate efficiently, with the experimental results showing a minimum efficiency gain of 1.74% across different operating scenarios.

6.2.4 Implementation of Reinforcement Learning on the UltraZohm Platform

RL, specifically the DDPG algorithm, is applied to optimize current control in PMSMs [20]. The DDPG agent adjusts voltage references to minimize errors in current tracking along the rotating d-axis and q-axis. This approach enhances the precision of the PMSM drive, which comprises a two-level voltage source inverter (VSI), by dynamically adapting control actions based on state measurements and derived quantities.

The DDPG-based control involves four neural networks: actor, critic, and their respective target networks. However, only the actor network is deployed for real-time control. Configurations vary depending on the reward functions and observation vectors used to evaluate performance across different setups. Training is performed using the MATLAB Reinforcement Learning Toolbox, with neural network structures tested for efficiency on AMD Threadripper processors. The training process involves extensive simulation episodes, and agents are optimized using hyperparameters derived from reinforcement learning literature.

The DDPG method is initially trained in a simulation environment, with the actor network, observation vector normalization, and output scaling being essential for real-time operation. The actor is implemented on the UltraZohm platform, which is an open-source rapid control prototyping system based on the Xilinx Zynq UltraScale + MPSoC. This platform features a combination of ARM processors (APU and RPU) and programmable logic (PL). It provides a robust environment for implementing complex control algorithms such as DDPG. In particular, the actor neural network, consisting of three hidden layers, is deployed in the PL to ensure the system meets real-time performance constraints. To ensure correctness and real-time capability before moving to physical hardware, Controller-in-the-Loop (CIL) testing is performed. In this step, the plant model is implemented in the PL of the UltraZohm, while the DDPG-CC controller operates in both the PS and PL. This intermediate testing process is crucial, because it allows for a thorough evaluation of the controller performance under realistic conditions without directly using physical equipment. The use of CIL significantly reduces implementation errors and makes the transition to real-world testing smoother. Furthermore, the UltraZohm platform and the DDPG testing procedures are integrated into an open-source project, enabling other researchers to easily reproduce the results. The availability of these resources allows anyone with a compatible Xilinx UltraScale + MPSoC device to recreate the experiment without needing access to specific PMSM or a full test bench. This promotes open science, making it easier to verify and extend the results while reducing the cost and resources typically needed for motor control experiments. This open approach to research supports reproducibility, fosters collaboration, and accelerates innovation in the field.

6.2.5 *Implementation of Autoencoder on FPGA*

The autoencoder framework is designed to evaluate degradation in railway traction inverters and handle multivariate time-series data, which is typical in condition monitoring [21]. By leveraging its recurrent structure, the model processes time-dependent data to monitor degradation online, providing robust and scalable assessments even under varying operating conditions. This capability makes it well-suited for deployment in complex and dynamic industrial settings, such as railway traction systems.

The proposed autoencoder-based method was implemented on an FPGA platform, specifically the Xilinx Zynq ZC706, which integrates a dual-core ARM Cortex-A9 processor system (PS) and an FPGA programmable logic (PL) unit within a single chip, as illustrated in Fig. 6.3.

This architecture allows time-critical functions to be allocated to the programmable logic of the FPGA for efficient processing, while the processor handles non-critical tasks.

Key functions, such as data normalization, neural network inference, and calculation of the MAE for the degradation index, are programmed in Verilog and executed on the PL, optimizing the system for real-time performance.

To facilitate real-time implementation, the MATLAB Deep Learning HDL Toolbox was employed. This tool generates HDL code from the autoencoder model, which is optimized for FPGA execution. The hardware setup processes real-time

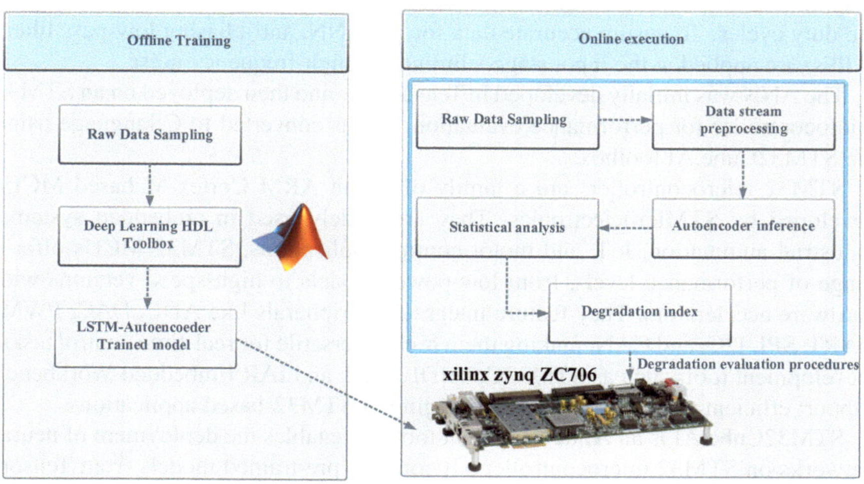

Fig. 6.3 Block diagram of the implementation of an autoencoder on an FPGA

multivariate data, performing preprocessing, inference, and degradation index calculations directly on the FPGA. This enables continuous monitoring of the inverter condition, utilizing less than 50% of the available FPGA resources, making the solution efficient and adaptable for real-time industrial applications.

6.2.6 Implementation of ANN in STEM

ANN has been applied to optimize the modulation of triple active bridge (TAB) converters [22]. These isolated multiport converters can support various loads and power levels, offering advantages in power density and efficiency. The ANN model addresses the complex optimization problem of minimizing the RMS current through the converter switches and transformer windings. Traditional optimization techniques are impractical for such systems due to the high number of modulation variables. Thus, the ANN model is trained to predict optimal duty-cycle values for modulating converter switches, thereby enhancing overall efficiency.

In this study, PLECS (piecewise linear electrical circuit simulation) was used for circuit simulation. PLECS is widely utilized for simulating power electronics and control systems, allowing for the analysis of current waveforms, voltage levels, and power parameters under different conditions. PLECS was employed to simulate the performance of the ANN-based modulation approach for various duty-cycle and operating conditions in the TAB converter.

The ANN-based modulation control was tested on a physical 5 kW TAB prototype setup. The ANN and converter control algorithms were implemented on an Imperix B-Box RCP controller, which drives six half-bridge modules. The ANN processes inputs from multiple sources, including voltages (V1, V2, V3) and power levels (P2, P3), at a rate of 10 kHz to balance computational demand and system responsiveness. PI controllers manage phase-shift adjustments, while the ANN focuses on optimizing the duty cycles. To ensure accurate data for the ANN, anti-aliasing low-pass filters (LPFs) are applied at the input stage, eliminating high-frequency noise.

The ANN was initially developed in TensorFlow and then deployed on an STM32 microcontroller for performance evaluation. It was converted to C language using the STM32Cube.AI toolbox.

STM32 microcontrollers are a family of 32-bit ARM Cortex-M based MCUs developed by STMicroelectronics. They are widely used in embedded systems, industrial automation, IoT, and motor control applications. STM32 MCUs offer a range of performance levels, from low-power models to high-speed versions with hardware accelerators. They feature integrated peripherals like ADC, DAC, PWM, UART, SPI, I2C, and CAN, making them highly versatile for real-time control tasks. Development tools such as STM32CubeIDE, Keil, and IAR Embedded Workbench support efficient programming and debugging of STM32-based applications.

STM32Cube.AI is an AI development tool that enables the deployment of neural networks on STM32 microcontrollers. It converts pre-trained models from TensorFlow, Keras, ONNX, and other frameworks into optimized C code for execution on

STM32 hardware. This tool provides model quantization, memory footprint optimization, and hardware acceleration support, allowing edge AI applications like predictive maintenance, object recognition, and motor anomaly detection to run efficiently on low-power STM32 MCUs. This makes STM32Cube.AI a powerful solution for integrating machine learning into embedded systems.

The ANN model comprises three layers with a total of 883 parameters, and it requires around 3.45 kB of ROM and 188 B of RAM. With 1463 multiply-and-accumulate (MAC) operations, the ANN achieves a computation time of 127.43 microseconds, ensuring efficient real-time performance. This compact ANN design provides reliable, optimized modulation control within resource constraints on low-cost microcontrollers, demonstrating feasibility for similar AI-driven applications in embedded power systems.

6.2.7 *Implementation of a Physical Informatics Neural Network on dSPACE MicroLabBox*

Physical informatics neural networks (PINNs) are implemented to enhance the control and modeling of power electronics systems, specifically in grid-forming microgrids [23]. The core advantage of PINNs lies in their ability to integrate physical laws directly into the training process, making them highly effective for modeling complex power systems. By incorporating both physics-based constraints and data-driven learning, PINNs ensure that the resulting model adheres to known physical behaviors while accurately capturing system dynamics under various uncertain operating conditions. This hybrid approach is particularly beneficial in power electronics, where both physical principles and operational data must be considered for optimal control.

Figure 6.4 illustrates the real-time process of the PINN. The system undergoes a structured process starting with simulation-based training. Initially, simulations generate the data required to train the neural network, which uses a combination of control-based and physics-based loss functions to minimize system errors. The network is then trained to replicate the behavior of the power electronics system (e.g., a DC/DC buck converter) and optimize the control strategies under various operating conditions. This training process ensures that the network can predict and adapt to changes in load, voltage, and other system parameters in real-world scenarios.

The experimental setup includes a programmable DC power supply, an electronic load, a DC/DC buck converter, and a dSPACE 1202 system. The control methodology is executed on the dSPACE platform, which generates PWM signals for the DC/DC buck converter ($C = 0.9\text{mF}$, $L = 1\text{mH}$). The Chroma programmable load is set to constant power mode to simulate constant power load (CPL) conditions. The implementation of the proposed method unfolds in three stages, as illustrated in Fig. 6.4. In the first stage, simulations generate data for the deep learning (DL) toolbox, where physics-based and control-based loss functions guide the predictions

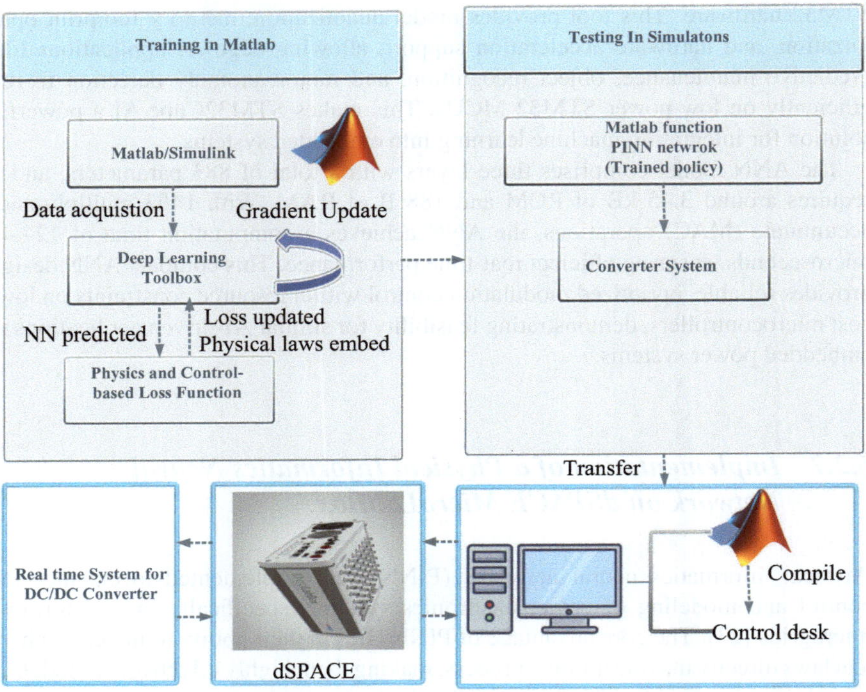

Fig. 6.4 Block diagram of the experiment implementation process of the PINN in MATLAB

and backpropagation of the neural network. This process effectively models the power electronics system and develops control laws through imitation learning. The second stage involves refining the neural network by adjusting the trained weights and biases, resulting in an enhanced network that embodies the optimal control policy. This refined network, which is embedded within a MATLAB function, serves as a dynamic tool to monitor parameters and improve control strategies using real-time data. The final stage involves converting the trained controller into C code, which is then compiled and integrated with the dSPACE MicroLabBox running on a real-time kernel for immediate deployment in power electronics systems.

6.2.8 Implementation of k-NN and RNN on FPGA

The authors of [24] explored experiment implementation of k-nearest neighbors (k-NN) and RNN to improve real-time device-level emulation of DC traction drive systems. These systems, which rely on high-frequency switching in power converters, require accurate modeling of switching transients to ensure reliable hardware-in-the-loop (HIL) simulation.

The k-NN module in the hybrid neural network is responsible for classifying the switching states of the power devices (e.g., IGBTs or diodes) based on input parameters like current, temperature, and switching status. It groups these states into predefined categories, which helps in selecting the appropriate model for transient prediction. Once the k-NN classification is performed, the RNN module is employed to predict the transient currents, which are associated with specific switching conditions. The RNN is trained using a set of samples categorized by current, temperature, and switching state, and its output provides a precise prediction of the transient behavior of the device. This hybrid AI system significantly reduces the complexity of traditional device-level simulations and provides real-time, accurate transient predictions, which are crucial for system-level analysis in DC traction drives.

Regarding hardware implementation, the AI-based model is deployed on an FPGA for real-time operation. A Virtex Ultrascale VCU118 evaluation board, which houses an XCVU9P-L2FLGA2104 FPGA, is used to implement both the k-NN classification and the RNN prediction modules. The large number of logic cells and DSP slices in the FPGA make it well-suited for the parallel processing required by both AI modules.

In the FPGA implementation of the k-NN classifier, the primary computational task is the calculation of the distances between the input sample and a set of training samples. The parallel processing capability of the FPGA is fully utilized in this phase, because multiple distance computations can be done simultaneously. The distance calculation typically involves straightforward operations such as subtraction, squaring, and summing the squared differences, which are well-suited for the parallel architecture of the FPGA. Each of these operations can be executed in parallel by different FPGA logic units, which dramatically reduces the time it takes to compute distances across a large set of training samples. After the distances are computed, the next step is sorting these distances to identify the k-nearest neighbors. The sorting step is essential for the classifier to determine which training samples are closest to the input. On the FPGA, the sorting operation is optimized to reduce its computational cost. A tournament-based sorting algorithm is employed, which allows for faster sorting with fewer clock cycles compared to traditional sorting algorithms. For example, sorting 128 elements can be completed in just 7 clock cycles, making the process highly efficient in hardware. The use of this specialized sorting algorithm ensures that the k-NN classifier can classify the switching states quickly, even when handling large datasets. Finally, once the distances are sorted, the majority voting process is used to classify the input sample. FPGA hardware can handle this step in parallel, reducing the time needed to determine the most likely classification.

In the FPGA implementation of the RNN, the trained weights of the network are stored in a look-up table, and the network performs a series of operations including multiplication, summation, and activation functions to produce predictions. The input to the RNN consists of classified switching states, which are passed through the network layers to generate the predicted transient current.

The computation of the RNN is broken down into layers: the input layer, the hidden layers, and the output layer. Each layer involves multiplying the input values by weights, adding biases, and passing the result through an activation function. On the FPGA, these operations are efficiently handled in parallel. The multiplication and summation steps for each layer are mapped to different processing units within the FPGA, allowing these operations to occur simultaneously across the entire network.

The activation function used in the RNN is implemented as a piecewise linear function, which simplifies the computation and reduces resource usage. This linear activation function allows the FPGA to compute results faster, meeting the stringent timing requirements of the system. For example, a typical RNN operation on a FPGA might involve a 140 ns delay for processing, which is well within the time-step requirements for real-time simulation (100 ns).

Both the k-NN and RNN modules benefit from the ability of the FPGA to handle multiple operations in parallel. This parallelism is key to achieving the real-time performance needed for hardware-in-the-loop (HIL) emulation. By distributing the computational tasks across the large array of logic cells and DSP slices of the FPGA, both classification and prediction tasks can be executed in parallel, significantly reducing the time needed for each step. For instance, the distance calculations, sorting, and majority voting steps of the k-NN classification all happen concurrently across multiple processing units, enabling the FPGA to classify switching states much faster than a sequential processor. Similarly, the layer-wise computations (multiplication, summation, and activation) of the RNN are parallelized to speed up prediction.

Moreover, the use of fixed-point arithmetic (59-bit format) is crucial for optimizing hardware resource consumption while maintaining high accuracy. Fixed-point representation reduces the complexity of the computations without sacrificing the precision required for transient prediction, further enhancing the efficiency of the FPGA implementation. The hybrid k-NN-RNN structure is trained using a combination of real-time and simulated data, including a wide range of current, temperature, and switching states. This training process uses a fixed-point representation (59 bits) to optimize the use of hardware resources while maintaining the required accuracy. The model is trained on a large dataset consisting of various device operating conditions, such as the IGBT collector current, diode forward current, and junction temperature, all of which are used to predict transient power losses. Once the model is trained, it can classify the switching state and predict transient currents in real-time by processing the input data through the k-NN and RNN modules on the FPGA.

6.2.9 Implementation of Random Forest on FPGA

Implementation the random forest (RF) algorithm on the FPGA for real-time fault detection and localization (FDL) in shipboard power systems (SPSs) transforms the traditional software-based approach into a hardware-accelerated solution [25]. The main advantage of using the FPGA for RF is its ability to execute multiple decision trees in parallel, thus dramatically speeding up the classification process, which is crucial for monitoring and ensuring the safety of marine vessels. This parallelism is essential for the efficient real-time monitoring of the SPS, because it allows the system to process complex feature sets and make quick fault detection decisions.

In the FPGA-based RF system, each decision tree, which is the core component of the RF algorithm, is processed in parallel. RF operates by aggregating the results from multiple decision trees, each trained on different subsets of data and features. The FPGA takes advantage of parallel hardware resources to evaluate each decision tree simultaneously. During classification, the input data is passed through each tree, and predictions are made in parallel for each tree. This parallel processing significantly reduces the time required to classify each sample and provides real-time evaluations of system conditions.

Each decision tree in the RF algorithm consists of nodes that compare input features against thresholds. In FPGA hardware, these nodes are mapped to logic units that perform feature comparisons. These comparisons occur in parallel, allowing the FPGA to traverse the decision tree quickly. Once a leaf node is reached, the FPGA retrieves the prediction for that tree. Multiple decision trees are evaluated simultaneously, and the results are aggregated using a majority voting mechanism. The parallelization of both tree evaluation and voting makes the FPGA-based implementation far more efficient than traditional software approaches, enabling rapid fault detection.

Memory management is a critical aspect of implementing RF on FPGAs. Decision trees, once trained, must be stored in the FPGA memory, along with the input data. Memory elements like Block RAM (BRAM) and external memory interfaces are used to store large datasets and trained model parameters. The FPGA needs efficient access to both training data and model parameters to make real-time predictions. To optimize this process, the FPGA architecture includes optimized memory management systems that ensure fast data access and minimal latency during classification.

To streamline the FPGA design process, high-level synthesis (HLS) tools are used. HLS allows designers to write code in high-level languages like C++ or OpenCL, which is then automatically converted into hardware description languages (HDL) like VHDL or Verilog. This approach accelerates development by automating low-level hardware design tasks, which allows designers to focus more on the structure of the algorithm rather than on the hardware intricacies. HLS also enables the FPGA to be optimized for parallelism, pipelining, and memory management, which are essential for the efficient execution of the RF algorithm.

The hardware implementation of the RF algorithm is carried out on the Xilinx Virtex UltraScale + VCU128 FPGA, which offers significant resources, including 2,852,000 programmable logic cells, a 340.9 Mb of Block RAM, and 9,024 DSP slices for high-speed mathematical operations. This FPGA is also equipped with a PCIe 3.0 interface, enabling high-speed communication with a host PC, where the simulation data from PSCAD/EMTDC is stored and transmitted to the FPGA for real-time classification. The FPGA processes this input data, evaluates the decision trees in parallel, and then aggregates the results to detect faults. These results are sent back to the host PC for further analysis and display.

References

1. Hussaini H, Yang T, Gao Y, Wang C, Urrutia M, Bozhko S (2024) Optimal droop control design using artificial intelligence techniques for electric power systems of more-electric aircraft. IEEE Trans Transp Electrification 10(1):2192–2206
2. Qu J, Zhang P, Jatskevich J (2024) Harmonic current optimization for torque ripple reduction in permanent magnet synchronous machine drives based on torque ripple surrogate model. IEEE Trans Power Electron 39(5):5108–5120
3. Qian Z et al (2023) Torque ripple reduction of PMSM based on modified DBN-DNN surrogate model. IEEE Trans Transp Electrification 9(2):2820–2829
4. Zhao S, Peng Y, Zhang Y, Wang H (2022) Parameter estimation of power electronic converters with physics-informed machine learning. IEEE Trans Power Electron 37(10):11567–11578
5. Karniadakis GE, Kevrekidis IG, Lu L et al (2021) Physics-informed machine learning. Nat Rev Phys 3:422–440. https://doi.org/10.1038/s42254-021-00314-5
6. Li X et al (2024) Temporal modeling for power converters with physics-in-architecture recurrent neural network. IEEE Trans Ind Electron 71(11):14111–14123
7. Sahoo S, Wang H, Blaabjerg F (2024) Uncertainty-aware artificial intelligence for gear fault diagnosis in motor drives. arXiv:2412.01272
8. Al-Kaf HAG, Lee JW, Lee KB (2024) Fault detection of NPC inverter based on ensemble machine learning methods. J Electr Eng Technol 19:285–295
9. Xiao D, Qin C, Ge J, Xia P, Huang Y, Liu C (2022) Self-attention-based adaptive remaining useful life prediction for IGBT with Monte Carlo dropout. Knowl-Based Syst 239:107902
10. Renkema Y, Brinkel N, Alskaif T (2024) Conformal prediction for stochastic decision-making of PV power in electricity markets. Electr Power Syst Res 234:Article 110750
11. Cifci A (2025) Interpretable prediction of a decentralized smart grid based on machine learning and explainable artificial intelligence. IEEE Access 13:36285–36305
12. Li S, Li T, Sun C, Yan R, Chen X (2023) Multilayer Grad-CAM: an effective tool towards explainable deep neural networks for intelligent fault diagnosis. J Manuf Syst 69:20–30
13. Li X, Lin F, Mantooth HA, Rodríguez-Andina JJ (2025) AI explainability for power electronics: from a Lipschitz continuity perspective. arXiv:2501.09948
14. Deng X, Feng X, Zhao S, Gou B, Wang H, Ge X (2024) Lightweight machine-learning-based diagnosis for power electronic systems subject to imbalanced data. IEEE J Emerg Sel Top Ind Electron 5(2):733–744
15. Xia Y, Xu Y (2021) A transferrable data-driven method for IGBT open-circuit fault diagnosis in three-phase inverters. IEEE Trans Power Electron 36(12):13478–13488
16. Naaz F, Herle A, Channegowda J, Raj A, Lakshminarayanan M (2021) A generative adversarial network-based synthetic data augmentation technique for battery condition evaluation. Int J Energy Res 45(13):19120–19135
17. Yao C et al (2024) Online open-circuit fault diagnosis for ANPC inverters using edge-based lightweight two-dimensional CNN. IEEE Trans Power Electron 39(4):3979–3984
18. Book G et al (2021) Transferring online reinforcement learning for electric motor control from simulation to real-world experiments. IEEE Open J Power Electron 2:187–201
19. Gou Y, Yu K, Wang F, Zhuo F (2024) Deep-learning-aided closed-loop synchronous rectification for isolated bidirectional DC/DC converter based on temperature gradient descent. IEEE J Emerg Sel Top Power Electron 12(1):66–81
20. Schindler T, Broghammer L, Karamanakos P, Dietz A, Kennel R (2023) Deep reinforcement learning current control of permanent magnet synchronous machines. In: 2023 IEEE international electric machines and drives conference (IEMDC), pp 1–7
21. Zhao S, Liu J, Chu K, Mu S, Wang H (2023) FPGA-based degradation evaluation for traction power module with deep recurrent autoencoder. In: 2023 IEEE energy conversion congress and exposition (ECCE), pp 1542–1548
22. Ibrahim AA, Zilio A, Younis T, Biadene D, Caldognetto T, Mattavelli P (2024) Optimal modulation of triple active bridge converters by an artificial-neural-network approach. IEEE Trans Ind Electron 71(3):2590–2600

23. Hui P, Cui C, Lin P, Ghias AM, Niu X, Zhang C (2024) On physics-informed neural network control for power electronics. arXiv preprint arXiv:2406.15787
24. Liang T, Huang Z, Dinavahi V (2020) Adaptive real-time hybrid neural network-based device-level modeling for DC traction HIL application. IEEE Access 8:69543–69556
25. Liu T, Liang, Dinavahi V (2020) Deep learning for hardware-based real-time fault detection and localization of all-electric ship MVDC power system. IEEE Open J Ind Appl 1:194–204

MIX
Papier aus verantwortungsvollen Quellen
Paper from responsible sources
FSC® C105338

If you have any concerns about our products,
you can contact us on
ProductSafety@springernature.com

In case Publisher is established outside the EU,
the EU authorized representative is:
Springer Nature Customer Service Center GmbH
Europaplatz 3, 69115 Heidelberg, Germany

Printed by Libri Plureos GmbH
in Hamburg, Germany